U.S. Marine Corps Combat Water Survival

Plus

Jungle Operations

Department of Defense

Pentagon
Publishing

MCRP 3-02C

Marine Combat Water Survival

U.S. Marine Corps

PCN 144 000069 00

To Our Readers

Changes: Readers of this publication are encouraged to submit suggestions and changes that will improve it. Recommendations may be sent directly to Commanding General, Marine Corps Combat Development Command, Doctrine Division (C 42), 3300 Russell Road, Suite 318A, Quantico, VA 22134-5021 or by fax to 703-784-2917 (DSN 278-2917) or by E-mail to **morgann@mccdc.usmc.mil**. Recommendations should include the following information:

- Location of change
  Publication number and title
  Current page number
  Paragraph number (if applicable)
  Line number
  Figure or table number (if applicable)
- Nature of change
  Add, delete
  Proposed new text, preferably double-spaced and typewritten
- Justification and/or source of change

Additional copies: A printed copy of this publication may be obtained from Marine Corps Logistics Base, Albany, GA 31704-5001, by following the instructions in MCBul 5600, *Marine Corps Doctrinal Publications Status.* An electronic copy may be obtained from the Doctrine Division, MCCDC, world wide web home page which is found at the following universal reference locator: **http://www.doctrine.usmc.mil**.

Unless otherwise stated, whenever the masculine gender is used, both men and women are included.

DEPARTMENT OF THE NAVY
Headquarters United States Marine Corps
Washington, D.C. 20380-1775

6 January 2003

FOREWORD

Marine Corps Reference Publication (MCRP) 3-02C, *Marine Combat Water Survival*, provides Marine Corps combat water survival techniques, procedures, and training standards. This publication also teaches Marines to cross water obstacles and perform water rescues correctly and safely.

This publication is the foundation for teaching Marines correct water survival techniques and procedures that are used throughout the Marine combat water survival program (MCWSP). Once an individual or a unit has completed the MCWSP, this publication can be used as a refresher course before water operations.

MCRP 3-02C supersedes Fleet Marine Force Reference Publication (FMFRP) 0-13, *Marine Combat Water Survival*, dated 16 September 1991.

Reviewed and approved this date.

BY DIRECTION OF THE COMMANDANT OF THE MARINE CORPS

EDWARD HANLON, JR.
Lieutenant General, U.S. Marine Corps
Commanding General
Marine Corps Combat Development Command

Publication Control Number: 144 000069 00

MCRP 3-02C Marine Combat Water Survival

Table of Contents

Chapter 2. Water Rescues

Chapter 3. Treatment of Casualties and Avoidance of Dangerous Marine Life

Chapter 4. Negotiating Water Obstacles

Chapter 5. Fording Waterways

Appendix. Knot Tying

Throughout history, water has posed special challenges to Marines and Sailors during times of both peace and war. Therefore, the inherent dangers associated with waterborne operations demand that Marines and Sailors receive proper water survival training. Combat units that are confident in their ability to work in and around water can use the water to their advantage in combat, and history is filled with examples of how the proper preparation or training for survival in water reduced or averted disaster.

USS *Indianapolis*

On Sunday, 29 July 29 1945, the heavy cruiser USS *Indianapolis* (the *Indy*) was en route to the Philippine Sea. Shortly before midnight (about 39 hours out of port), the *Indy*, running blacked out and unescorted, was rocked by two explosions on her starboard side. With communications smashed, the ship could not signal its distress and sank within 15 minutes.

Three life rafts and a floater net supported a few survivors, but the rest of the survivors drifted about, held up by rubber life belts or Mae Wests. About 60 seamen died the first night. Survivors assumed the ship would be reported overdue in Leyte, and that they would be rescued within 2 days. By Monday evening, however, panic began to set in as some life jackets lost their buoyancy from the long immersion. Some men even fought over life jackets, which resulted in at least 25 deaths. No one dared sleep for fear of losing his jacket.

Throughout the next several days, in-transit aircraft flew nearby without spotting the desperate seamen. As best they could, the men kept together, some tying long ropes to each other, floating like corks on a net.

Not until late Thursday morning, 3 1/2 days after the ship sank, were the men discovered. Luckily, a plane on a routine flight over the area spotted the survivors. When ships picked them up that night, the survivors learned they had never been reported overdue. Every one of the 1,196-man crew was a casualty; 880 crewmembers were listed as dead or missing. Although many lives were lost, the innovative and expedient use of flotation devices and float techniques employed by the survivors helped save hundreds of lives.

USS *America*

On Thanksgiving eve, 23 November 1995, the USS *America* made its way through the Arabian Sea. Twenty-year-old, Marine Lance Corporal (LCpl) Zachary Mayo was unable to sleep and, wanting some fresh air, made his way onto an open-air platform near the aircraft hangar bay, which was three levels below the sleeping quarters. While he was on the platform, the ship veered suddenly, throwing LCpl Mayo through the platform's protective bars and into the sea, 30 feet below.

Frantic, LCpl Mayo called out in vain to the watchmen on the flight deck, which was 64 feet above him. It soon became clear to him that the USS *America* would keep its course into the Gulf of Oman until his absence was discovered at morning muster. The LCpl took a moment to consider his situation. Since land was at least 100 miles away, swimming was suicide; he would have to stay afloat until a search party found him.

Using the techniques he had learned during combat water survival training, LCpl Mayo made a flotation device out of his coveralls and tried to relax. Meanwhile, business continued as usual aboard the USS *America*. Since LCpl Mayo was on special assignment with the hazardous materials division, his absence went unnoticed until a petty officer asked several Sailors if they had seen their shipmate recently. By the time a roll call had been completed, LCpl Mayo had been adrift at sea for over 24 hours. Although three, fixed-wing Viking aircraft were deployed to search for LCpl Mayo, most people aboard ship feared the worst.

After 34 hours at sea, LCpl Mayo was discovered by fishermen on a Pakistani fishing boat. LCpl Mayo's survival is a testament not only to his incredible physical courage, but also to the soundness of the lifesaving training and techniques he received during combat water survival training.

Chapter 1
Survival at Sea

As a Marine, you face a variety of potential water emergencies whenever you cross expanses of water: ships, watercraft, and amphibious assault vehicles (AAVs) can sink; aircraft can crash into the sea; or you can accidentally fall into the water. However, there are some basic precautionary measures you can take to protect your safety and reduce your chances of becoming a water casualty. Determine the following information as soon as you board any type of vessel. Your knowing the following information may save your life or the life of your fellow shipmates.

- How many life preservers and lifeboats/rafts are on board?
- Where are the life preservers and lifeboats/rafts located?
- What type of unit survival equipment is on board?
- Are individual survival kits issued to each person on board?
- How much food, water, and medicine do the survival kits contain? When was the last time the contents were inspected for proper quantities and shelf life expiration?
- Is there sufficient survival equipment available for the number of personnel?
- How many other personnel are there on board, and where are they located?
- What are the egress procedures for the ship, boat, watercraft, AAV, or aircraft?

Abandoning Ship

When you embark on a Navy ship, you will receive abandoning ship instructions from Navy personnel. If given the order to abandon ship, report to your designated assembly area and put on a life preserver. DO NOT inflate the life preserver until you are clear of the ship. Torn life preservers will not inflate and inflated life preservers can block you, and those behind you, from exiting the ship. A flotation device that has been inflated may also burst if you jump from a significant height. See pages 1-15 through 1-35 for staying afloat with and without a life preserver.

DO NOT remove your clothing, boots, or shoes before abandoning ship. Your trousers and blouse may be the only flotation devices available if your life preserver is faulty or becomes damaged, and your clothes can provide some insulation from the cold water. However, remove your soft cover and place it in a cargo pocket for later use. The soft cover is both lightweight and a good protection against sunburn caused by the sun's rays reflecting off the water.

Jettisoning Equipment

Equipment should be kept properly packed and waterproofed in case you have to abandon ship. If entering the water from a height greater than 30 feet, wearing your equipment (e.g., pack, helmet, gas mask) could cause injury. Upon impact with the water, the helmet will "cup" air inside of it. The chin strap may also create a "hanging effect" as you submerge from the force of the fall. Therefore, you should remove your helmet and gas mask before abandoning ship.

If you are unable to maintain buoyancy due to the amount of equipment secured to your pack and body, then jettisoning some of your equipment may become necessary. Equipment that you should always retain includes canteens of freshwater, first aid kit, soft cover, and survival kit. The survival kit should include first aid items, water purification tablets or drops, fire starting equipment, signalling items (e.g., flashlight, strobe light, chemlights), food procurement items, and shelter items. Other items in a survival kit include sunburn lotion and lip balm, knife, goggles/sunglasses, plastic bag, matches and lighter, mirror. See Marine Corps Reference Publication (MCRP) 3-02F, *Survival*, for a detailed list of survival items and applications.

Abandoning Ship Technique

When abandoning ship, safety considerations must be observed. Use the following technique when abandoning ship without your combat gear:

Place your hands on their opposite shoulders, forming a criss-cross pattern.

Step to the edge of the ship's deck and check the water below for debris or survivors. If the water is clear, look straight ahead and prepare to jump. If the water is not clear, move to another location.

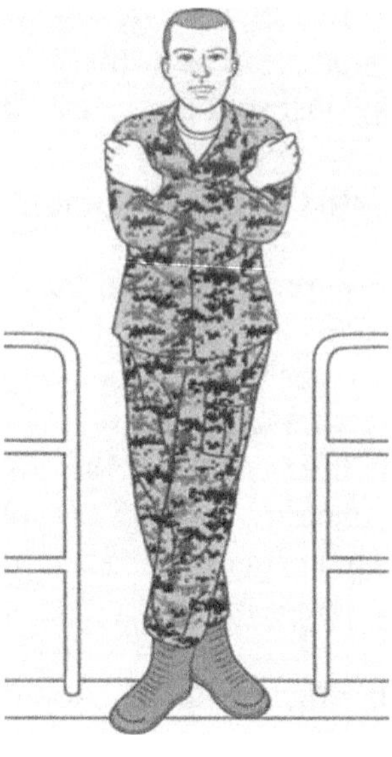

NOTE: DO NOT hold your nose as you abandon ship. If you do hold your nose, the force of impact into the water could jar your arm and hand and cause you to break your nose.

Step off the side of the ship with a smooth, 30-inch stride. DO NOT DIVE OFF THE SHIP. DO NOT LOOK DOWN AT THE WATER. LOOK STRAIGHT AHEAD. Looking down at the water can render you unconscious or cause injuries upon impact.

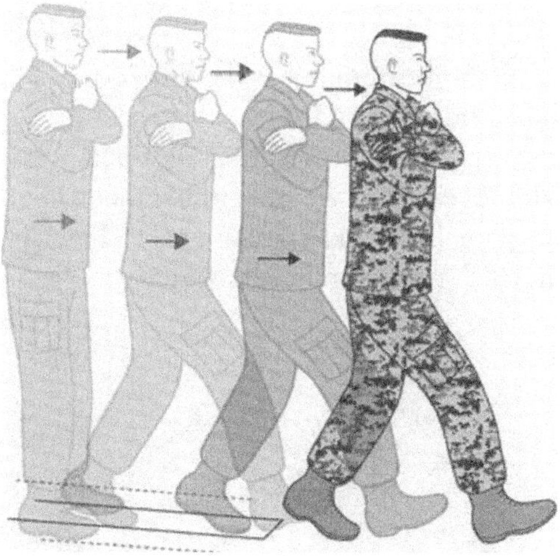

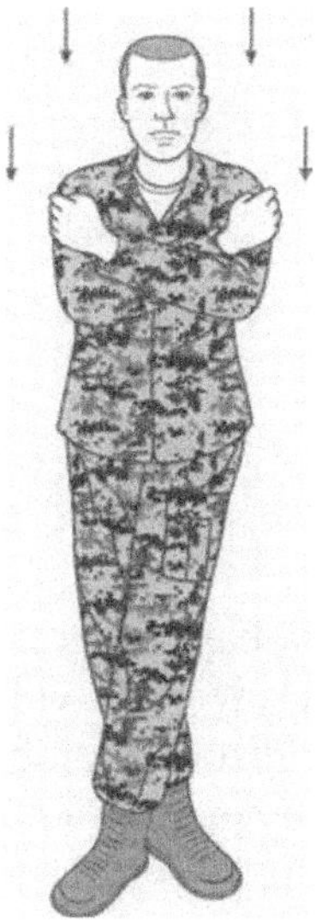

Bring your trailing leg forward during the fall. Cross your trailing leg behind your leading leg.

Keep your head parallel to the water's surface until hitting the water.

You should remain in the abandon ship position until your descent into the water has almost stopped. However, the weight imbalances in your body may cause you to be in a "J" shape under the water

.

Once your downward motion has ceased, your feet may be parallel with the ocean bottom or you may be nearly inverted with your feet over your head. To counteract potential disorientation, you should pause briefly and allow the natural buoyancy of your torso to bring your body to a nearly upright position.

Floating debris can cause hazards. Therefore, you should swim upward, extending one arm (hand is shaped as a fist) upward to feel for obstructions. If you encounter debris, try to push it away or surface in a different location.

Swim away from the ship. DO NOT LOOK BACK AT THE SHIP. Looking back could slow your movement away from the area. Remember, your objective is to leave the area as quickly as possible because—

- Equipment and debris may be falling from or spilling out of the ship.
- Additional casualties can occur if individuals abandoning the ship fall on top of swimmers already in the water.
- Swimmers close to the sinking ship may get pulled underneath the water by the suctioning effect of the ship as it goes under.

Modified Abandoning Ship Technique

When abandoning ship while wearing full combat gear (weapon, helmet, and a properly waterproofed pack), safety considerations must be observed. The modified abandoning ship technique is used while wearing full combat gear and exiting from a height less than 30 feet.

WARNING

When exiting from a height greater than 30 feet, remove your helmet and pack. Fasten the helmet to the pack or place it inside the pack before jettisoning. If jettisoning gear from the ship or aircraft, check the water below for survivors before throwing the gear forward of the intended jump area. Once in the water, retrieve your gear and swim out of the area.

Place your weapon over one shoulder, muzzle down, with the weapon parallel to your side.

Place your arm and hand along the weapon and hold it to your side.

Take your free hand and place it on top of your helmet to prevent neck/spinal injury from the force of the water pulling upward on the helmet as your body enters the water.

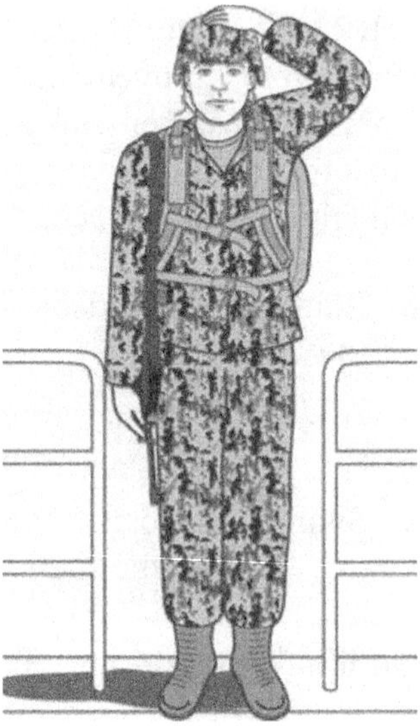

Note: An alternate method is to place your weapon over one shoulder, muzzle down, with the weapon parallel to your side. Reach across your body and grasp the sling of your weapon and hold it to your body. Take your free hand and place it on top of your helmet.

Step to the edge of the platform and check the water below for debris and survivors. If the water is clear, look straight ahead and prepare to jump. If the water is not clear, move to another location.

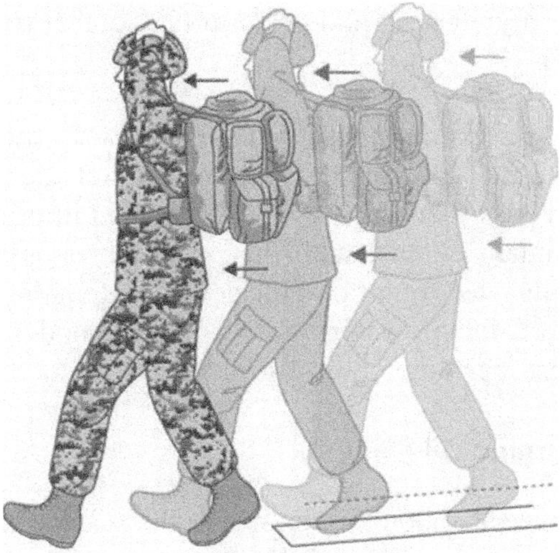

Step off the side of the platform with a smooth, 30-inch stride. DO NOT DIVE OFF THE SHIP, DO NOT LOOK DOWN AT THE WATER, LOOK STRAIGHT AHEAD. Looking down at the water can render you unconscious or cause injury upon impact.

Bring your trailing leg forward during the fall. Cross your trailing leg behind your leading leg.

Keep your head parallel to the water's surface until hitting the water.

You should remain in the modified abandoning ship position until your descent into the water has almost stopped. The buoyancy of a properly waterproofed pack will immediately pull you to the surface. Once you break the water's surface, unsling your weapon and loop the sling over your head. You may have to seesaw the sling so as to ensure the sling passes between your helmet and pack. Once over your head, loop the sling around your neck with the weapon aligned in the center of your body, muzzle down. Lean back on your pack and perform the combat travel stroke to exit the area.

> *Note:* You may also remove the pack (unsnap the quick-release strap on one side of the pack, and pass the strap to your other hand in order to maintain contact with the pack, which is functioning as your flotation device). Remain horizontal in the water with your pack under your chest. Perform a breast stroke kick.

Surface Burning Oil Swim

After you have abandoned ship, rise to the surface using the techniques shown on the page 1-6. However, you must remember that fuel from sinking ships or downed aircraft will float on the surface of the water. Therefore, you must move clear of the floating fuel by swimming away from the ship or aircraft as soon as possible. Either swim upwind (into the wind) of the ship/aircraft or swim against the current. Either method allows you to move away from the fuel and the wind/current will push the fuel past you. To properly execute a surface burning oil swim—

Extend your arms overhead as far as possible.

Wave your arms back and forth vigorously to splash a hole while moving upward.

Splash as long as possible to push burning fuel away from the surfacing area.

Use your arms and hands to sweep away fuel and debris.

Kick your legs in a constant breast stroke kick.

Extend your arms (palms outward) forward on the surface, arms shoulder-width apart.

Pull your hands in and back toward the chest.

Stop your hands in front of your face and rotate them so that your palms face forward (roughly halfway out of the water).

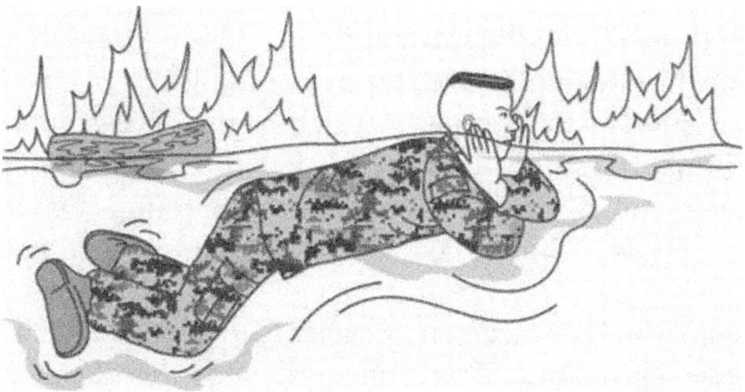

Sweep your arms forward to a full extension at the shoulder width. This splashes debris, oil, or burning liquids aside. To reduce the chance of fatigue, use two short splashes to the front to extend the path.

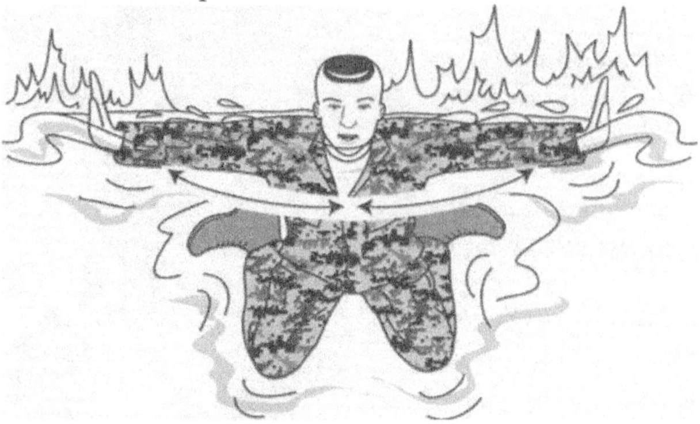

Repeat the preceding step as necessary while swimming clear of the area.

Surviving With a Pack

If packed properly, your pack will float, and it is your key piece of equipment for staying afloat and overcoming water obstacles. If the pack's contents are properly waterproofed, it can support you (with a combat load) in the water. Buoyed up by a waterproofed pack, you eventually emerge from the water with all your equipment (e.g., boots, helmet, flak jacket, weapon, survival items).

Your pack floats based on a scientific principle known as Archimedes' principle. This principle states that an object submerged in a liquid is buoyed up by a force equal to the weight of the liquid displaced (pushed aside) by the object. If the weight of the displaced liquid is greater than the weight of the object, the object floats. If the weight of the displaced liquid is less than the weight of the object, the object sinks. For example, a machine gun sinks in the water, but it still weighs less in the water than it does on land. Even though a machine gun sinks, it is still buoyed up by a force equal to the weight of the water it displaces. For this reason, you should not try to hold yourself or your equipment any higher out of the water than they would naturally float; doing so wastes both energy and body heat.

Preparing Equipment

Before packing your pack, you must prepare your gear/equipment. Tape or pad all sharp edges and equipment corners. Ideally, your gear/equipment is placed in plastic bags and the plastic bag is then placed inside the standard-issued, rubberized, waterproof bag. Waterproof bags are not completely water tight; henceforth, the added protection of first wrapping the gear/equipment in a plastic bag, then placing the plastic bag inside the waterproof bag. Large plastic bags (e.g., trash bags) work well for bulky equipment (e.g., sleeping bags, field jackets, shelter halves, gas masks). Small plastic bags work best for small items (e.g., shaving gear).

> *NOTE:* If a gas mask must be carried outside the pack, cover it with a waterproof bag.

You will need the following items, which are available through the supply system, to prepare your equipment for packing:

- Waterproof bags:
  Pistol bag, plastic size #1 (8" x 18").
  Rifle bag, plastic size #2 (10" x 56").
  Machine gun bag, plastic size #3 (15" x 56").
  Multipurpose bag, cover, plastic size #4 (20" x 84").

- Panama work vest (life preserver).
- Small plastic bag, self sealing (6" x 6" and 12" x 12").
- Riggers tape, olive drab (2 1/2").
- Willie peter bag, waterproof clothing.

Tying Waterproof/Plastic Bags

Try to remove the excess air from the waterproof/plastic bag before securing the bag's opening. If filled with air, the bag can burst if pressed from the outside. Perform the following steps to tie a waterproof/plastic bag:

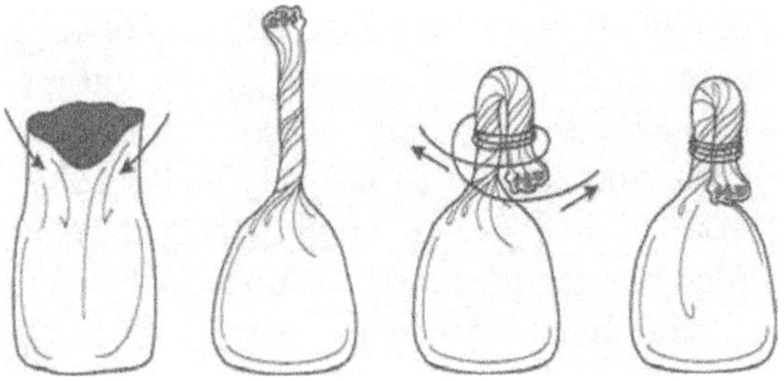

Packing the Pack

Place filled and tied bags inside the pack. Carefully handle sharp items (e.g., tent pegs, poles) to prevent puncturing the bag. Place items in the pack in order of expected use. Close the pack and its compartments. Attach sleeping mats or bags as high as possible on the outside of the pack.

Swimming With the Pack

A useful technique for propelling yourself forward while wearing a pack is the combat travel stroke. This technique allows you to float nearly horizontal and to propel yourself forward with bicycle-style kicks and breast stroke-style arm sweeps. To execute the combat travel stroke—

Body position
Keep the upper part of your body prone to the water, let your legs dangle horizontally, and keep your face up.

Arm action
Extend your hands out in front of your waist. Sweep your arms slightly downward and back 90 degrees to propel your body through the water. Move your hands back to the front of your waist. Repeat.

Leg action
Continuously move your legs in a bicycle-like movement, bringing your knees up high and step out.

Breathing
Keep your face out of the water during the stroke, and breathe freely.

Staying Afloat With a Life Preserver

The best form of flotation is to find any kind of floating object that will keep you and your equipment out of the water or minimize your exposure to the water. Life preservers are the best method, as they allow you to wear your clothes for heat retention and sunburn prevention. Marines use two basic classes of life preservers: inherently buoyant life preservers and inflatable life preservers.

Inherently Buoyant Life Preservers

Inherently buoyant life preservers are either vest-type (worn like a jacket) or yoke-type (worn around the neck). The preserver's outer envelope is either a cotton or water resistant material that encloses a removable fibrous glass or plastic foam filling.

In the Marine Corps, the most common type of inherently buoyant life preserver is the vest-type with collar, known as the kapock preserver. The kapock consists of collar straps, upper front chest straps, leg straps, and waist drawstrings that secure the preserver to you. The leg straps, which are fitted on both sides of the life preserver, ensure that the preserver remains around your chest while you are in the water. A chest strap is attached to the life preserver to facilitate lifting you out of the water. The strap can also be attached to other survivors or to lifeboats to reduce the fatigue that results from holding onto a floating/secured object by hand.

Inflatable Life Preservers

Marine Corps aircraft and AAVs have inflatable life preservers on board. The inflatable life preserver issued to Marines is known as the LPP (life preserver personal). The LPP is capable of both oral inflation and CO_2 cartridge inflation. The LPP consists of buoyancy chambers, CO_2 inflator, and an oral inflation tube. The buoyancy chamber is deck gray in color and is made from a neoprene-coated nylon fabric.

Inflatable life preservers must be stored in a cool, dry place. Heat, moisture, and light cause deterioration of the life preserver material. Do not stow CO_2 cylinders near steam lines or radiators. Heat can increase the pressure inside the cylinders causing them to explode. Avoid sharp edges in stowage. Sharp edges increase wear and tear on the life preservers and may also puncture inflatable buoyancy chambers.

> **CAUTION:** Do not inflate the life preserver until you are clear of the aircraft, ship, or vehicle. Torn life preservers will not inflate and inflated life preservers can block you, and those behind you, from exiting the aircraft, ship, or vehicle.

To don and adjust the LPP—

Remove the LPP from its storage container.

Fasten the belt fasteners in front with the pouch in the rear.

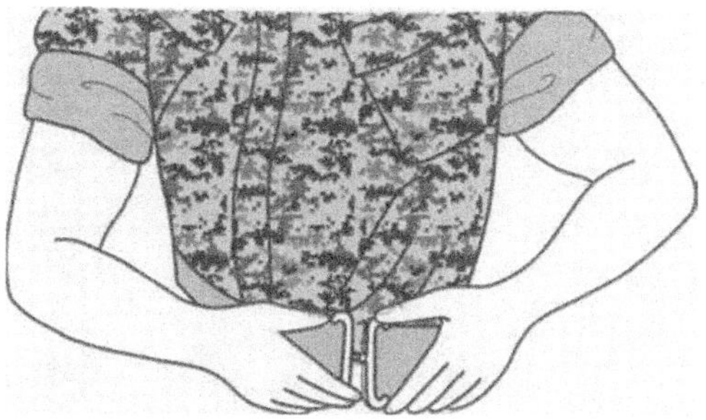

Adjust the belt to fit; secure any excess belt by mating the hook and pile tape.

Rotate the pouch to the front and re-adjust the belt if necessary.

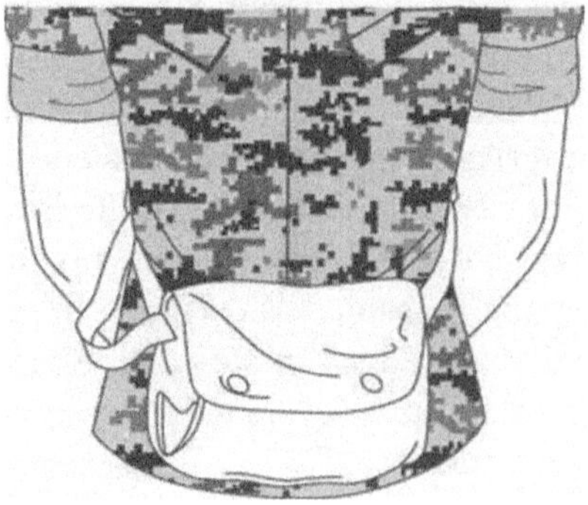

Open the snap fasteners on the pouch and unfold the life preserver.

Place the deflated life preserver over your head.

Place the storage container into the pouch after donning the life preserver.

Lift the lower end of the life preserver out of the pouch.

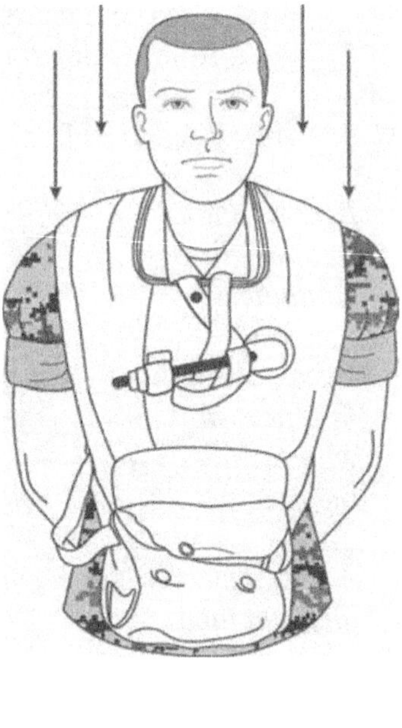

Inflate the life preserver by pulling on the lanyard attached to the CO_2 inflation valve or by blowing on the end of the oral inflation valve.

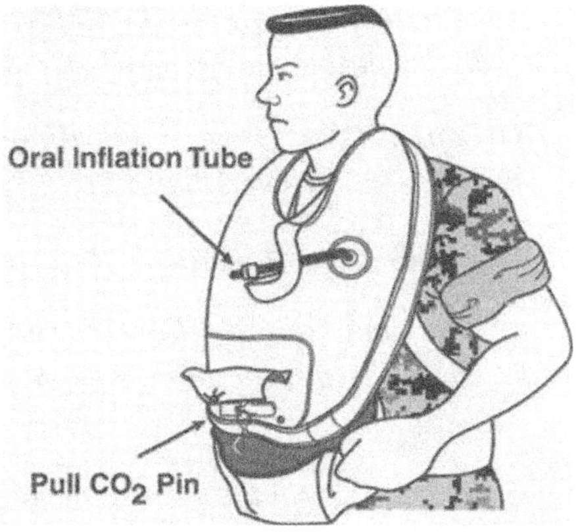

Staying Afloat Without a Life Preserver

You may be in open water without any floating objects or a life preserver to help you survive. If so, uniform blouses and trousers can be made into expedient flotation devices.

Floating With an Inflated Blouse

It is possible to float by a bubble of air trapped in the shoulders of your blouse. The air rises to the back and shoulders of the blouse and supports you at the water's surface. An inflated blouse is also a temporary flotation device used by weaker swimmers while trying to remove their trousers. There is a primary and an alternate way to create a bubble of trapped air in a blouse—

Primary Method
Turn the collar inside the blouse to help create a seal.

Unbutton top button and pull collar around mouth and nose.

Take a deep breath and bend forward slightly at the waist. Exhale one-half to three-quarters of a breath into the blouse.

Grasp and twist the collar with one hand to create a seal, this prevents air from escaping out from the collar.

Use your free hand and feet to stroke and kick to the surface.

Gather and hold the blouse tightly at the collar and stomach level to prevent the blouse from losing air if it floats up too high.

Splash water on the blouse periodically to prevent the material from drying, dry material allows air to escape.

Repeat inflation as required.

Alternate Method
Turn the collar inside the blouse to help create a seal.

Unbutton the second button from the top.

Take a deep breath and bend forward slightly at the waist.

Place your mouth and nose inside the hole created by the open button and exhale one-half to three-quarters of a breath into the blouse.

Grasp material at the unbutton portion and pull downward.

Use your free hand and feet to stroke and kick to the surface.

Splash water on the blouse periodically to prevent the material from drying, dry material allows air to escape.

Repeat inflation as required.

Floating With Inflated Trousers

In warm water, trousers can be used as a primary expedient flotation device. However, in cold water, submerging your head to remove and inflate your trousers results in heat and energy losses that negate the benefit of using the trousers as a flotation device. Once your trousers are inflated, you float motionlessly as if wearing a life preserver. If needed, assume the heat escape lessening posture (known as the HELP position, see page 1-37) to slow heat loss. As trousers dry, air leaks out of the legs. To slow this process, occasionally splash water on the fabric. Reinflate trousers as needed.

Sling Method

The sling method works if you are a strong swimmer or naturally very buoyant. Take the following steps to inflate trousers using the sling method:

Take a deep breath, bend over, and remove your boots.

NOTE: Retain your boots. Tie the boot laces together and suspend the boots from your blouse or hang them around your neck so that they rest on your chest.

Remove your trousers. Button or zip the trouser's fly closed. This allows you to control airflow.

Tie the bottoms of the trouser legs in a square knot. (The appendix illustrates various knots.)

Ensure that the front (fly) of the trousers faces you.

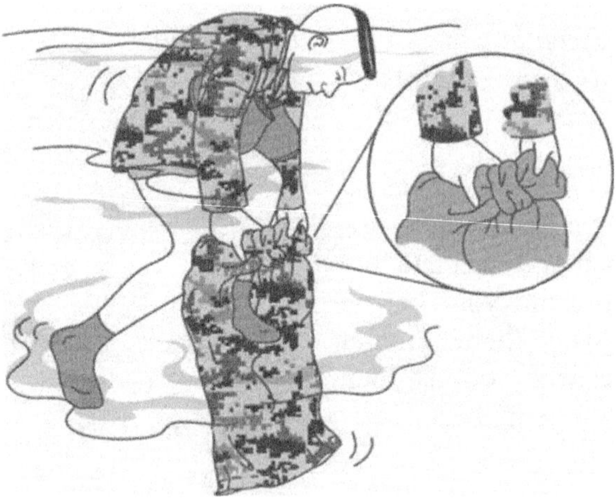

Hold the trousers above the water's surface and behind your head. Grasp both sides of the waistband and open with both hands.

Kick strongly to stay on top of the water while slinging the trousers overhead in order to trap air into them.

Once the waistband is submerged in the water, air is trapped in the legs.

Hold and seal the waistband underwater.

Slip the inflated legs over your head. Hold the waistband in toward your chest, the fly facing your body. To prevent air from escaping from the trousers, seal the waistband by either folding it or twisting it.

Lie back and relax, resting the back of your neck against the knot.

Splash water on the trousers periodically to prevent the material from drying. Dry material allows air to escape.

To replenish air in the trousers, you will use a technique known as the scooping method. With one hand on the open waistband, extend the trousers in front of you just below the surface of the water and scoop air bubbles with your free hand into the open waistband until the trousers have sufficient air. Repeat as necessary.

Splash Method

The splash method is an alternative to the sling method. As with the sling method, you must kick strongly to remain at the surface. To inflate trousers using the splash method, perform the following:

Take a deep breath, bend over, and remove your boots.

Note: Retain your boots. Tie the boot laces together and suspend the boots from your blouse or hang them around your neck so that they rest on your chest.

Remove your trousers. Button or zip the trouser fly closed. This allows you to control airflow.

Tie the bottoms of the trouser legs in a square knot. (The appendix illustrates various knots.)

Ensure that the front (fly) of the trousers faces you.

Hold the trousers at the water's surface out in front of you by the waistband with the fly up.

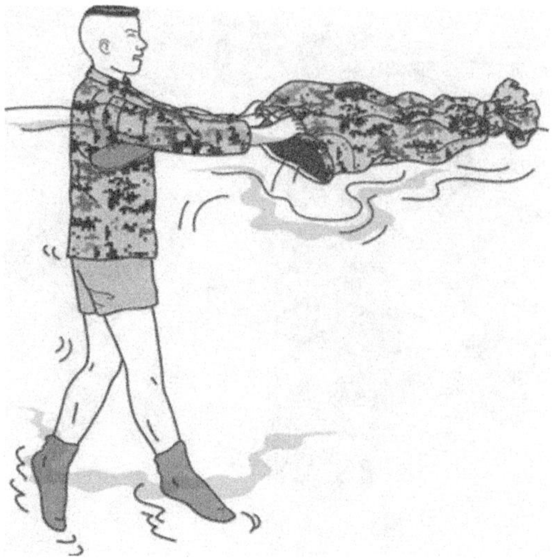

Grasp the waistband at the surface with one hand. Insert your free hand into the waistband, palm down.

Flutter your hand rapidly to create bubbles. This sends a mixture of water and air bubbles into the trousers. The water passes through the fabric. The air remains trapped in the legs.

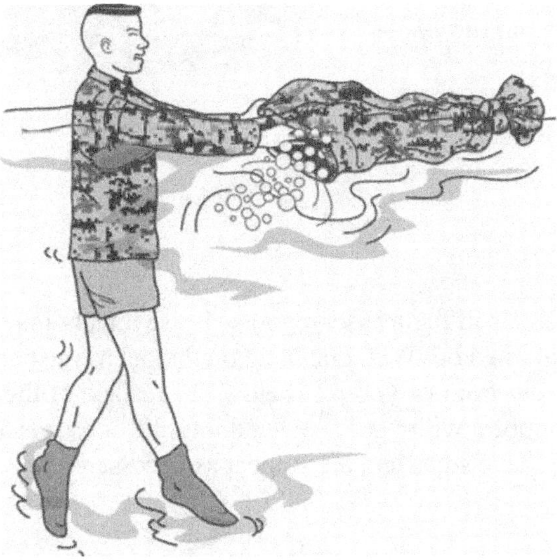

Hold and seal the waistband underwater.

Slip the inflated legs over your head. Hold the waistband in toward your chest, the fly facing your body. To prevent air from escaping from the trousers, seal the waistband by either folding it or twisting it.

Lie back and relax, resting the back of your neck against the knot.

Splash water on the trousers periodically to prevent the material from drying. Dry material allows air to escape.

To replenish air in the trousers, you will use a technique known as the scooping method. With one hand on the open waistband, extend the trousers in front of you just below the surface of the water and scoop air bubbles with your free hand into the open waistband until the trousers have sufficient air. Repeat as necessary.

Blow Method

The blow method is an alternative to the sling method. Use the blow method if you are a weak swimmer. Take the following steps to inflate trousers using the blow method:

Take a deep breath, bend over, and remove your boots.

NOTE: Retain your boots. Tie the boot laces together and suspend the boots from your blouse or hang them around your neck so that they rest on your chest.

Remove your trousers. Button or zip the trouser fly closed. This allows you to control airflow.

Tie the bottoms of the trouser legs in a square knot. (The appendix illustrates various knots.)

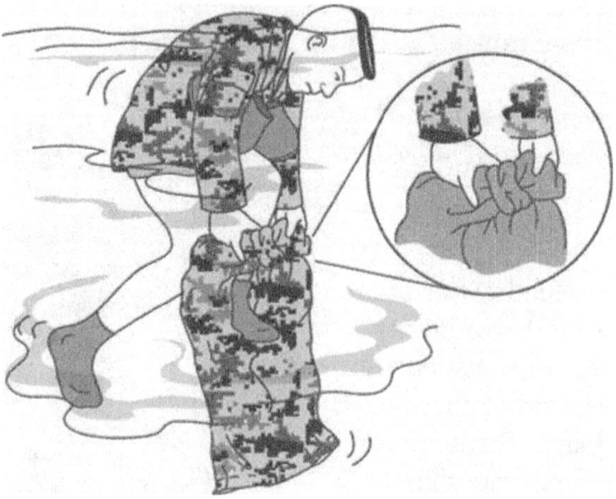

Ensure that the front (fly) of the trousers faces you.

Hold the trousers at the water's surface. Grasp both sides of the waistband and open with both hands.

Take a deep breath.

Drop 2 feet below the water's surface, pulling the waistband underwater.

Hold the waistband open with both hands and blow air into the trousers.

To fill the trousers with air, surface while keeping the waistband underwater, breathe in again, drop below the water's surface, and blow air into the trousers. Repeat these steps until the trousers are filled sufficiently. Once trousers are filled—

Hold the waistband underwater. Twist and pinch it off.

Slip the inflated legs over your head. Hold the waistband in toward your chest, the fly facing your body. To prevent air from escaping from the trousers, seal the waistband by either folding it or twisting it.

Lie back and relax, resting the back of your neck against the knot.

Splash water on the trousers periodically, to prevent the material from drying. Dry material allows air to escape.

To replenish air in the trousers, you will use a technique referred to as the scooping method. With one hand on the open waistband, extend the trousers in front of you just below the surface of the water and scoop air bubbles with your free hand into the open waist-band until the trousers have sufficient air. Repeat as necessary.

Avoiding Heat Loss in Cold Water

The rate of heat exchange in the water is about 25 times greater than it is in air of the same temperature. When you are immersed in cold water, hypothermia occurs rapidly due to the decreased insulating quality of wet clothing and as a result of water displacing the layer of still air that normally surrounds the body. You also lose about 50 percent of your body heat through your head; therefore, keep your head out of the water. Other areas of high heat loss are the neck, the armpits/sides, and the groin.

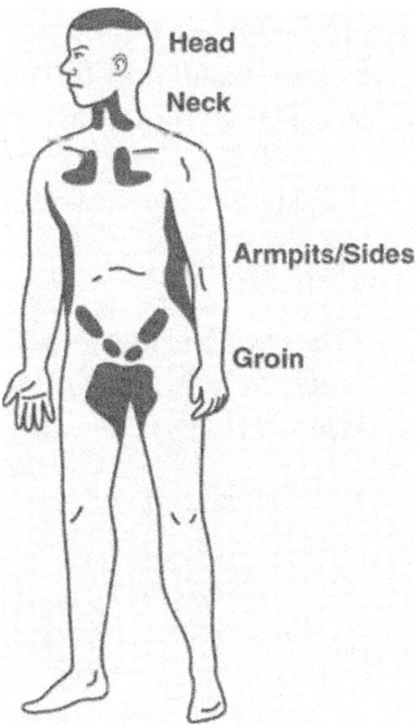

In cold water, DO NOT SWIM TO STAY WARM. Swimming, even with a slow and steady stroke, produces a lot of heat that is lost in the water. The heat loss can produce hypothermia that slows body functions and can result in serious injury or death. Remaining motionless conserves body heat three times longer than swimming. SWIM only if you have a flotation device and the shoreline is visible.

Individual Protection From the Cold

If you are equipped with a life preserver, assume the heat escape lessening posture (known as the HELP position) to slow heat loss and to protect major blood vessels near the body's surface. These areas lack insulating fat and are vulnerable to the chilling effects of cold water. To assume the HELP position—

Tuck your chin down tightly to cover your throat.

Draw your legs up in a fetal position to protect the groin.

Place your arms across your chest, tuck your hands into your armpits.

Wear some type of head covering (e.g., stowed cover, towel, handkerchief) to lessen heat loss through the scalp if head covering is available.

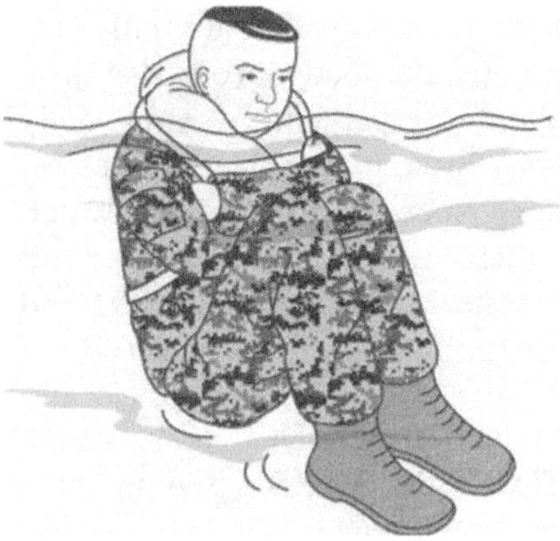

Group Protection From the Cold

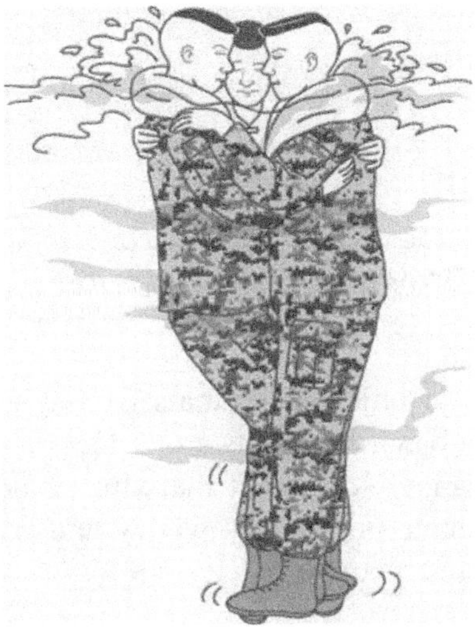

If three or more Marines are in the water and are equipped with life preservers, they should wedge tightly together and lock arms to form a circle known as a huddle position. This position protects vulnerable areas from heat loss.

A casualty who is suffering from the effects of the cold can be placed within the huddle to be surrounded by warmer water. If in the water for a prolonged period, it is recommended that everyone be rotated inside the huddle to maintain or re-warm each person's internal core temperature. If there are more than five Marines, they should make clusters of huddle positions.

Contact with other swimmers provides survival advantages—

Creates a larger target for search and rescue aircraft.

Provides additional warmth in cold water.

Improves morale.

Re-establishes the chain of command.

Reduces shock and panic.

Provides opportunities to administer first aid.

Supports exhausted Marines.

Drownproofing Methods

An object that floats on the surface has positive buoyancy. An object that floats a few feet beneath the surface has neutral buoyancy. An object that sinks has negative buoyancy. Most people have positive buoyancy and will float at the water's surface. To test your buoyancy—

Stand in water that is at your chest level or deeper.

Take a full breath.

Bend forward slowly.

Relax and wait.

If you have positive buoyancy, you will slowly rise to the surface. If you have neutral buoyancy, you will float a few feet beneath the surface. If you have negative buoyancy, you will sink. Regardless of whether you naturally float or sink, you can remain at the surface for extended periods without a life preserver if you exercise the appropriate drownproofing methods or survival strokes, which are based on your buoyancy.

Drownproofing methods consist of the T-method and the sweep. The breast stroke, side stroke, and elementary backstroke are the most common survival strokes. With any drownproofing method or survival stroke, remember the acronym **SAFE**:

SLOW EASY MOVEMENTS
Move slowly to conserve energy and minimize heat loss.

APPLY NATURAL BUOYANCY
Use natural buoyancy to support the body.

FULL LUNG INFLATION
Fill the lungs with each breath. Do not hold air in the cheeks.

EXTREME RELAXATION
Tight muscles are denser than relaxed ones and do not float as well.

Crawl Stroke

The crawl stroke, sometimes called the front crawl or free style, is the fastest stroke. To execute the crawl stroke—

Body position
Lie horizontal, on your stomach, in the water.

Look forward and downward at a 45 degree angle with the waterline between your eyebrows and hairline. Your head position is important as it assists in cutting a path through the

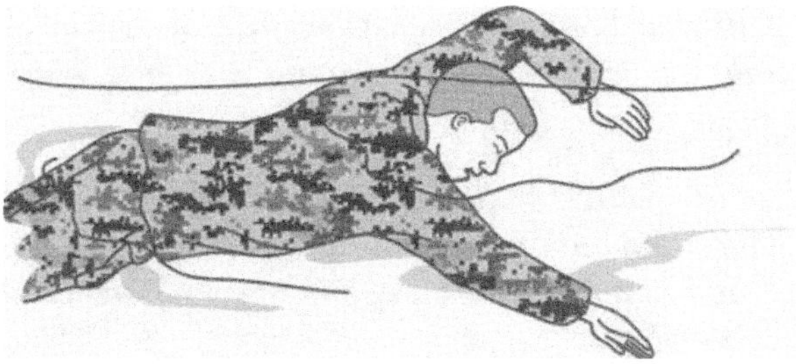

water. If your head is too high in the water, your lower body sinks significantly, making your stroke less efficient. If your head is too low in the water, water washes over your shoulders and neck, causing unnecessary drag.

Arm action

Arm action occurs in three phases: catch, propulsion, and recovery.

Fully extend one arm forward of your body, this positions your hand to catch the water in preparation for the propulsion phase.

To catch the water, bend your wrist (with your palm pointing outboard) and make an "S" shape (or inverted "S" shape) with your hand, ensuring that your hand does not cross the center of your body. Your left hand makes an "S" shape and your right hand makes an inverted "S" shape with your hand finishing at shoulder level.

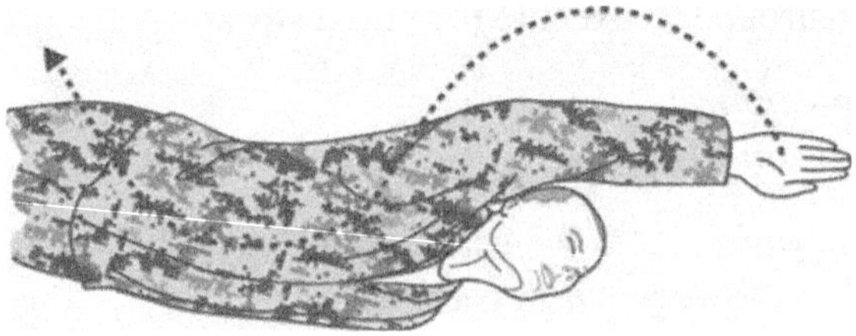

Push with your hands in a rearward fashion toward your feet until your arm is fully extended along your side, keeping your hands close to your body.

To begin the recovery phase, bend your arm at the elbow and raise your hand out of the water. Your hand breaks the surface of the water and maintains a height of 2 to 3 inches above the water's surface.

With your hand and arm moving in a forward manner, bring your hand past your head until your arm is about three-quarters of the way extended. At this point, turn your hand with your palm outboard so as to allow your thumb and fore-finger to enter the water first.

Once your hand enters the water, continue to push your arm forward until it is fully extended.

You are now prepared to catch the water again. These steps are performed in an alternating pattern: when one arm is catching and propelling, the other arm is recovering.

Leg action
Use a flutter kick to create the leg action for the crawl stroke. This kick is used for both propulsion and keeping the lower body horizontal with the water's surface. The flutter kick is an alternating leg action: one leg is kicking in a downward motion while the other leg is recovering to the surface to prepare for the next kick. The size of the kick ranges from 12 to 15 inches and depends on your height.

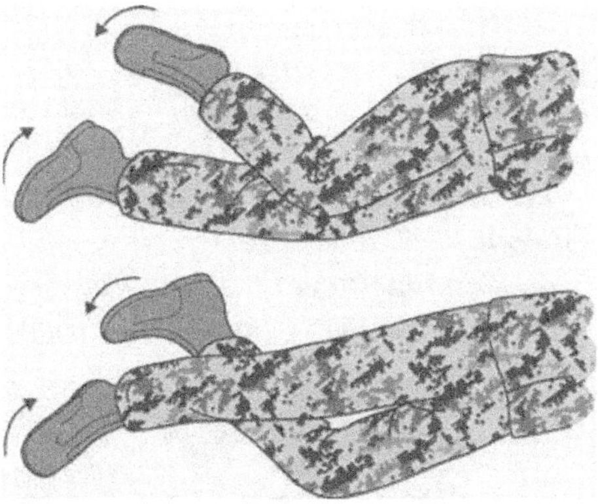

Maintain your legs in a semi-rigid manner.

Generate power for the kick from the hips.

Keep your feet loose at the ankles, they trail behind your legs and act as "flippers."

Execute the propulsion phase of this kick with a downward thrust of your leg.

Execute the recovery phase by pushing your leg back to the surface. This phase is complete when your foot reaches the surface.

Breathing
Breathe during the recovery phase of the arm action.

Roll your body and rotate your head to the side of your body where the arm recovery is occurring, this rolls the water away from your mouth. Keep your chin pushed back toward your shoulder.

Exhale while your face is still submerged and inhale when your face breaks the surface.

Breathe either bilaterally or rhythmically. To breathe rhythmically, breathe on the same side of your body every time your arm cycle occurs. Bilateral breathing requires you to breathe every one and a half arm cycles. To breathe bilaterally, breathe when your right arm is recovering, your face goes back in the water and the next time you breathe is when your left arm is starting its recovery three arm strokes later.

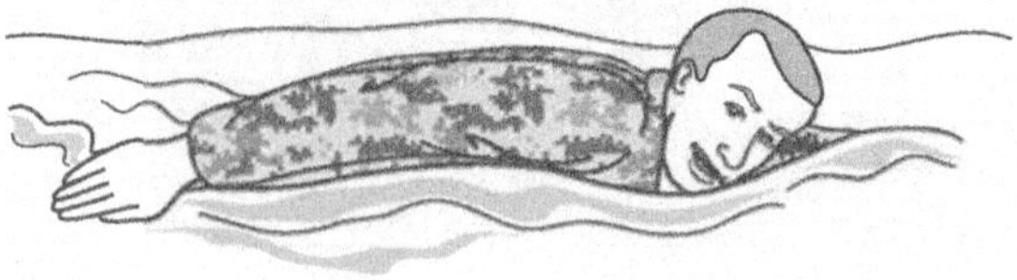

Bilateral breathing is the preferred method for breathing; it prevents the chance of hyperventilation and allows your body to maintain a lateral position in the direction you are swimming.

Coordination
This stroke uses constant arm and leg action.

T-Method

The T-method is a basic drownproofing method. This is the best survival technique if you have negative buoyancy. To execute the T-method—

With your face out of the water, take a deep breath and submerge your face in the water while holding your breath.

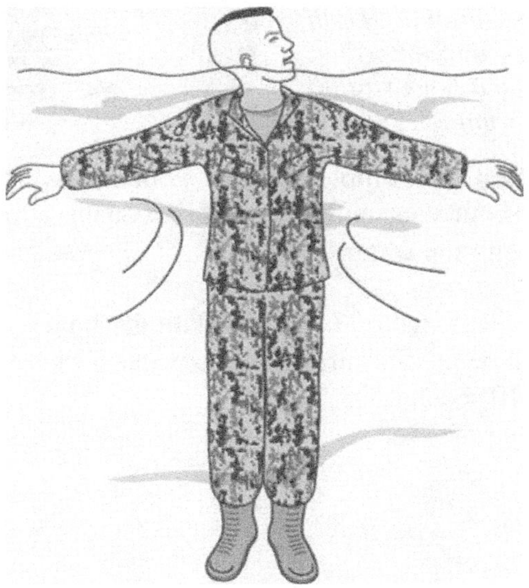

Float with your body in a horizontal position, arms extended from your side, and legs extended and joined.

Move your hands up to your armpits by tracing an imaginary line along your ribs.

Extend your arms outward (horizontal) to your sides, your body position resembles the letter "T."

Step out and forward with one leg and point your other leg to the rear, your knees should be slightly bent. Simultaneously, bring your arms down to your sides. Then bring your legs back together. You complete the steps by exhaling most of your air and preparing to surface your face to obtain another breath of air. Hold your head out of the water. Tilt your head back slightly. Breathe normally.

Once your breath is complete, move your hands up and down directly in front of your body. Do this two or three times to slow your descent into the water.

> NOTE: To avoid hyperventilating, hold your breath below the surface of the water for no more than 10 seconds.

The Sweep

The sweep works well if you have slight to excellent positive buoyancy. To execute the sweep—

Float face down in the water: bend 45 degrees at the waist, arms and legs dangling, head hanging down, relax all muscles.

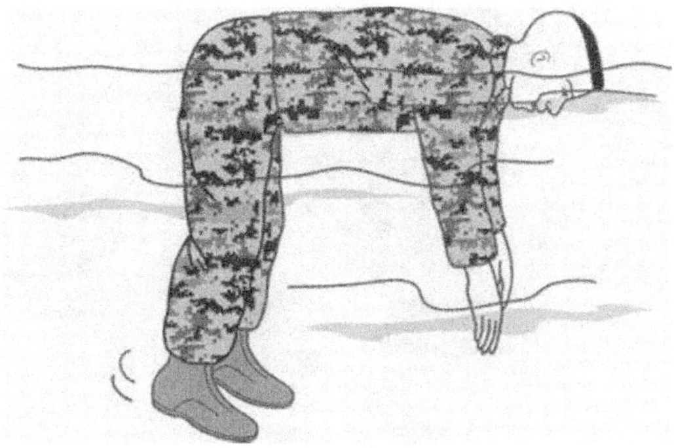

Spread your feet slowly to prepare for a single kick (one leg is forward and one leg is rearward).

Cross your arms in front of your chest, palms outboard with the back of each hand touching the opposite ear.

Exhale prior to raising your head for a breath.

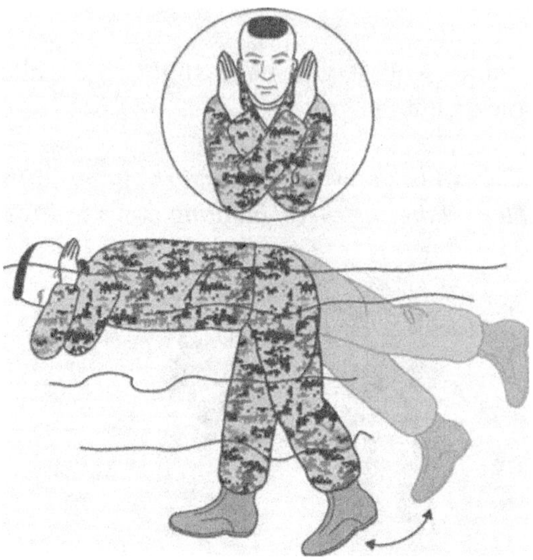

Bring your legs together and sweep your arms down and out until the arms are fully extended out to the sides. This raises your face above the water and allows you to catch a breath of air.

With your air supply replenished, return your face to the water and relax while sweeping your arms in a downward motion in front of your body to prevent/slow your descent. Do not hold your breath for more than 10 seconds. More than 10 seconds enhances your chances of shallow water black out and subsequent drowning.

Breast Stroke

Use this stroke to swim underwater, through oil or debris, and in rough seas. If you are a good swimmer and not wearing combat gear, the breast stroke is the best stroke for long-range swimming because it provides good visibility and allows you to conserve your energy and maintain a reasonable speed.

Body position

Lie prone in the water. Swim with your trunk and legs projecting back and down at an angle of 20 to 30 degrees.

Extend arms out in front (hands together [side by side]), and extend legs behind (toes pointed) to prevent drag.

Face downward, looking forward at a 45 degree angle to break the water and to prevent water from washing into the collar area causing drag.

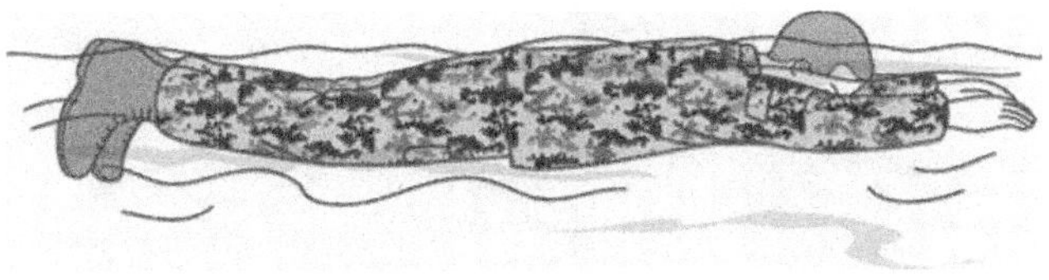

This is known as a glide.

Arm action

Turn your palms outward and bend your arms slightly.

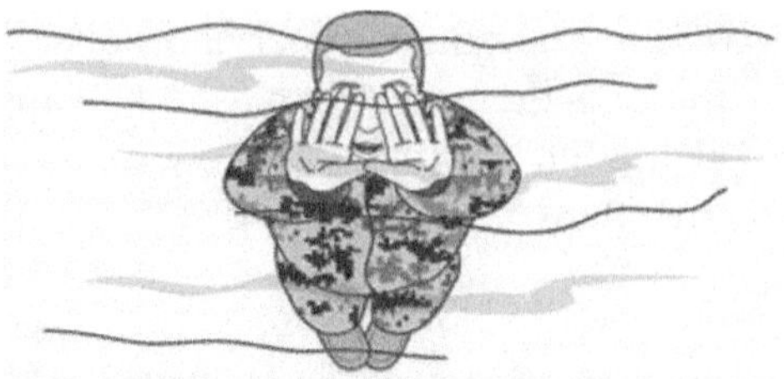

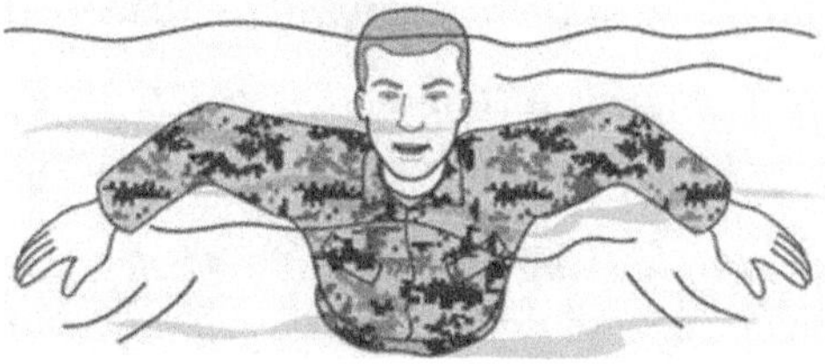

Sweep your arms sideward and slightly downward until your hands are opposite and slightly below your shoulders.

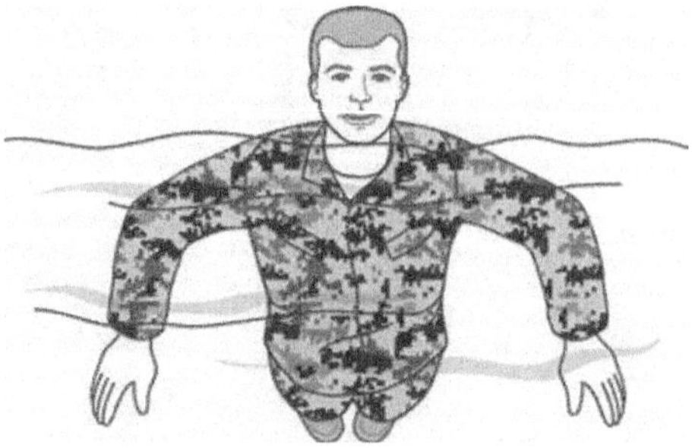

Rotate your head up, breathe once your mouth breaks the surface.

Bring your hands and arms up along your chest and thrust them forward until they are extended and ready to execute the next arm pull.

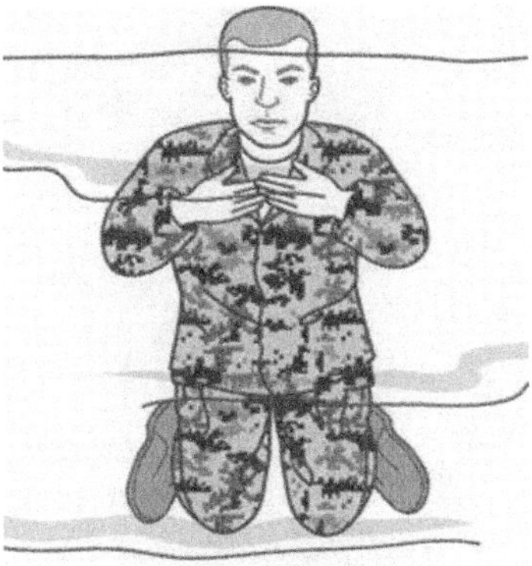

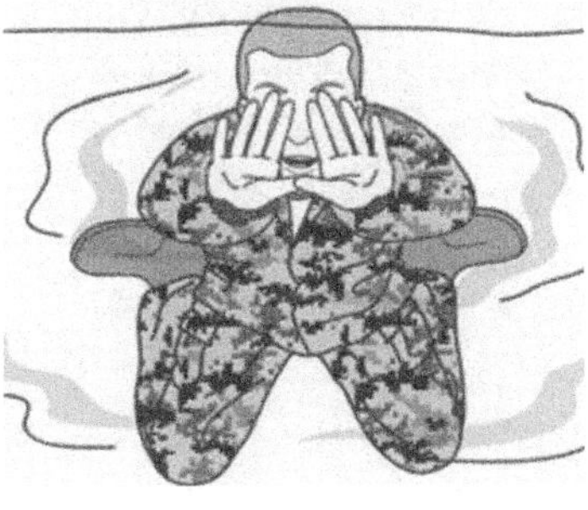

As the arms start their recovery into the glide, the head should rotate forward, resubmerging the face.

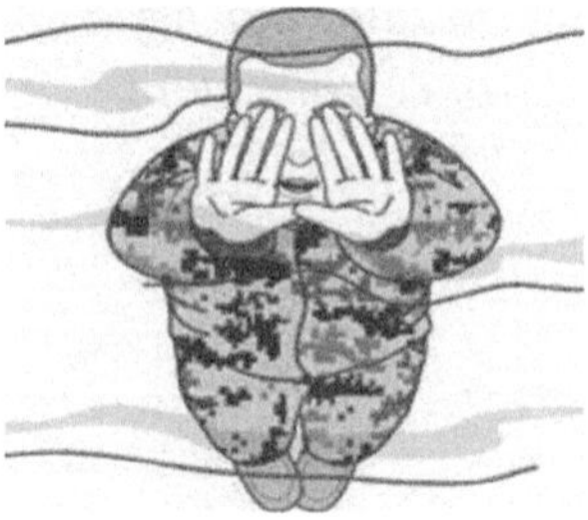

Leg action

Draw your heels toward your buttocks, establish a 45 degree bend in the knees.

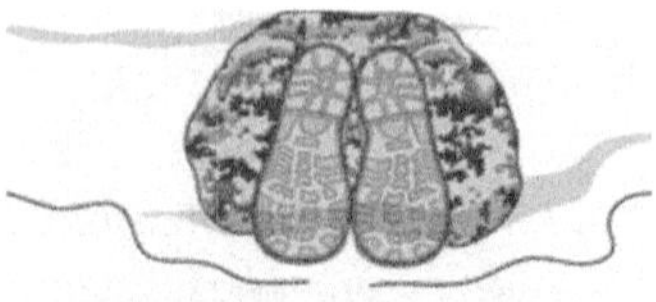

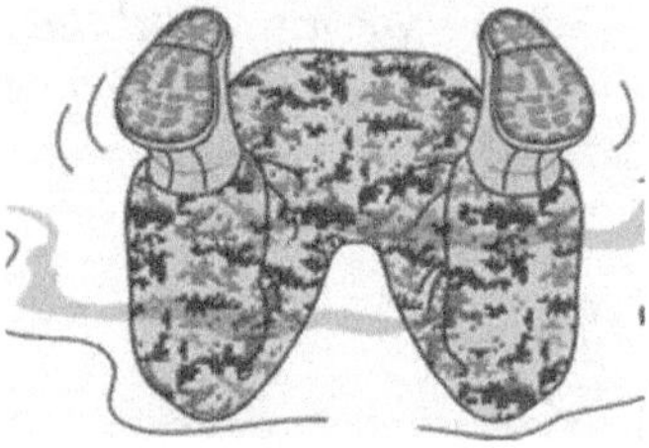

Thrust your legs outward and rearward, then squeeze them together. The whipping action of the feet aids forward propulsion.

This is known as the breast stroke kick.

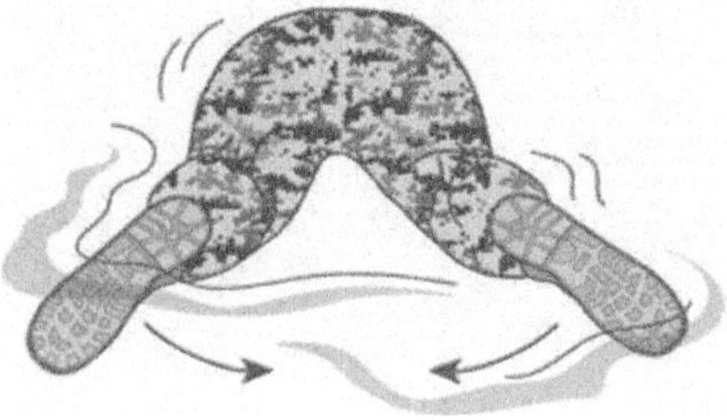

Breathing
Inhale during the arm pull and exhale through your mouth and nose during the finish of the breast stroke kick and glide.

Coordination
The stroke movement is in three counts:

Begin your arm pull. Near the finish of the pull, flex your knees and bring your heels toward your buttocks. The arm pull counteracts the resistance created by the knees.

As the arm pull is completed, thrust your hands forward, kick your legs outward and rearward, and squeeze them together.

Glide through the water for approximately 1 to 3 seconds or until your forward momentum decreases, then begin the next stroke.

Side Stroke

The side stroke is a survival stroke because you use both arms for buoyancy, with each arm creating a slight propulsion. The majority of your body's propulsion comes from your kick. To execute the side stroke—

Body position
Lie on your side with your lead (bottom) arm extended beyond (with a slight bend in your elbow) your head and in line with your body. Palm is down and your hand is submerged 6 to 8 inches.

Extend your trail (top) arm down the length of your body over your thigh.

Keep your legs straight and together, toes pointed rearward.

Keep your face out of the water, this allows for free breathing.

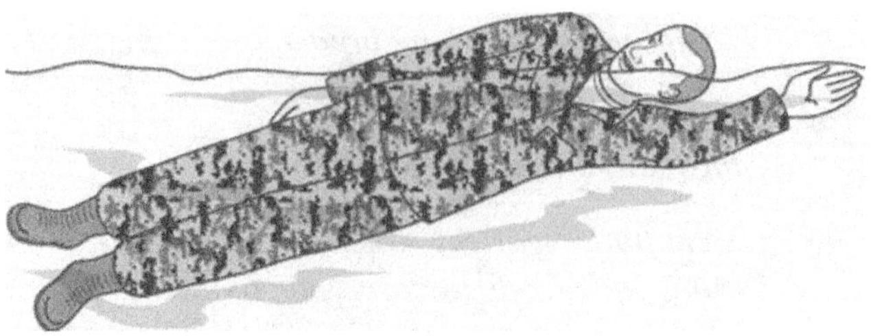

This is known as the glide.

Arm action
With your lead arm, pull your arm downward, while flexing at the elbow, until it is straight down from your shoulder.

Rotate your shoulder and pull your elbow into your side. This should put your lead hand at shoulder level. At the same time, turn your palm toward your face and thrust forward to your original, extended position.

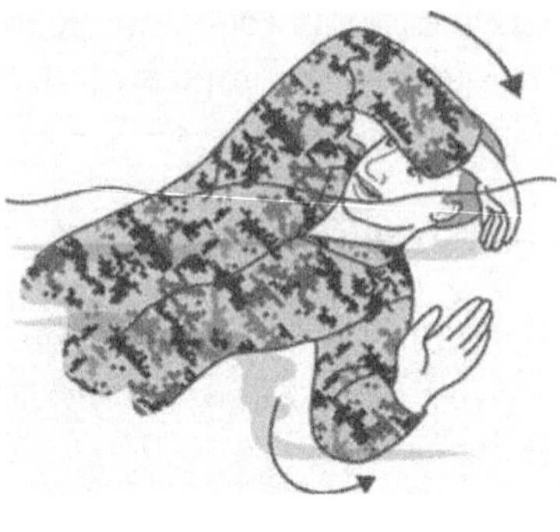

Draw your right hand upward in front of your chest to shoulder level. Rotate your palm toward your feet, then push it downward in front of your body toward your feet to catch the water.

Push your trail hand backward to its original position on top of your thigh. (Your trail hand starts forward and meets your lead hand at your chest/shoulder.)

Leg action

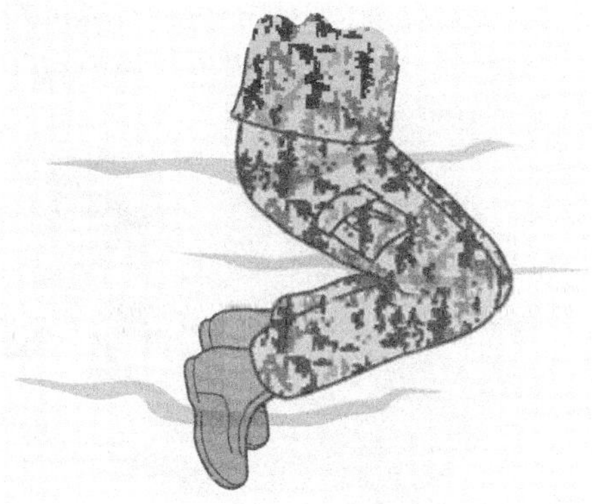

To perform the scissor kick, the top leg always goes forward and the bottom leg always goes rearward. From the extended position, draw—or recover—your feet toward your buttocks until your legs are bent at a 45 degree angle at the knees and the hips are flexed at a 45 degree angle with the thighs.

Once the legs have completed their recovery and while maintaining a 45 degree bend in the knees, extend the legs fully into a "V" shape in order to catch the water for the propulsion phase.

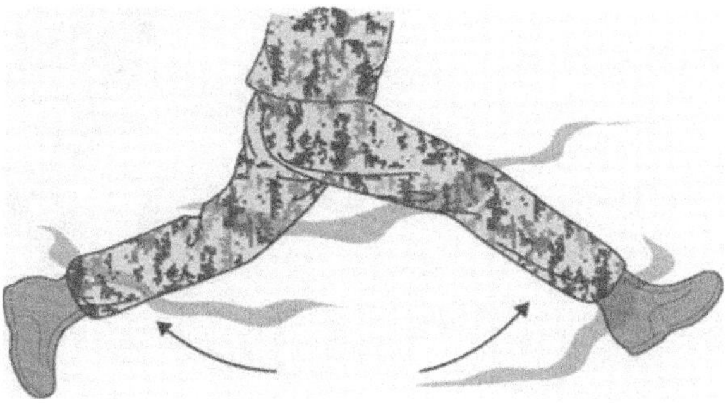

Once the legs are separated and extended forward and rearward to the "V" position, sweep the legs together until the feet are together.

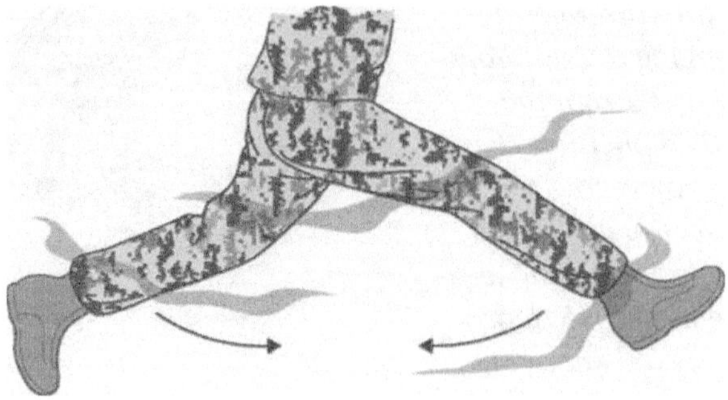

You are now in the glide position.

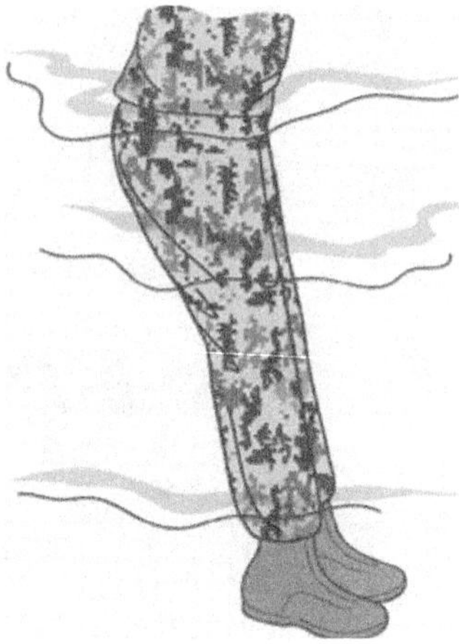

Breathing

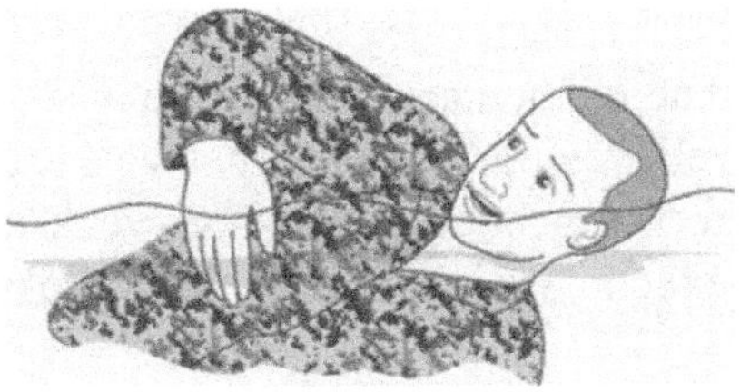

As long as your face remains clear of the water, it is a free breathing stroke. However, it is recommended that you exhale then quickly inhale when the legs are sweeping back together in the scissor kick. This is when the body reaches its highest point in the water, thus clearing the face completely from the surface of the water making it the optimum time to breathe.

Coordination

Begin the stroke with the downward pull of your lead arm. At the same time, bring your trail arm upward and draw your knees up to begin the kick. Let the thrust of the lead arm, push of your trail arm, and the kick of your legs coincide in order to finish the glide position. Glide through the water for approximately 1 to 3 seconds or until your forward momentum decreases, then begin the next stroke.

Elementary Backstroke

The elementary backstroke is also an excellent survival stroke. It relieves the muscles that you use for other strokes, and it is the recommended stroke for weak swimmers or nonswimmers. To execute the elementary backstroke—

Body position
Start on your back.

Face up, chest up, and hips up, keeping an arch in your lower back with arms pressed to your sides and your legs extended and joined to prevent drag.

Arm action
Trace your hands up your sides to an area near your armpits then extend your arms out to the sides to form the letter "T" (palms facing feet), locking out the elbows.

> NOTE: Don't raise your arms above your head. This creates drag, changes your body position, and submerges the head.

Slap your palms to your thighs using a strong sweeping motion.

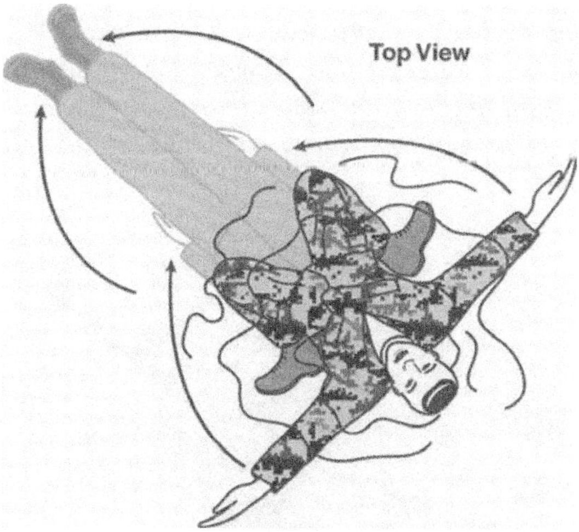

Leg action
Bend both legs at the knee (90 degree angle) slightly separating your knees and drawing your heels downward to a point under and outside your knees. The knees are spread as wide as the hips or slightly wider depending on the body type of the swimmer.

Circle around in a whipping action, ending with legs in a glide position.

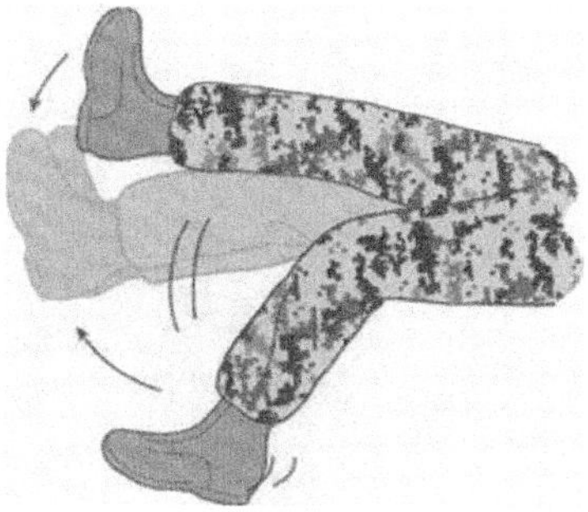

Breathing

Breathe anytime during this stroke. However, it is recommended that you exhale then quickly inhale when your arms are sweeping back toward your sides and while your legs are sweeping back together. This is when the body reaches its highest point in the water, thus clearing the face completely from the surface of the water and making it the optimum time to breathe.

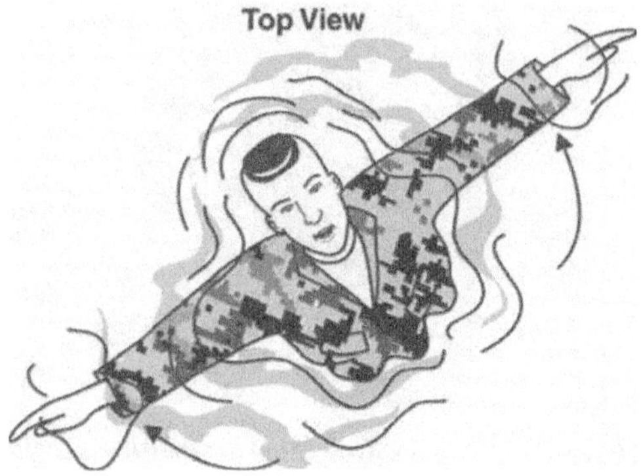

Coordination

The stroke movement occurs in three counts (recovery, catch, power).

Begin the arm pull (recovery).

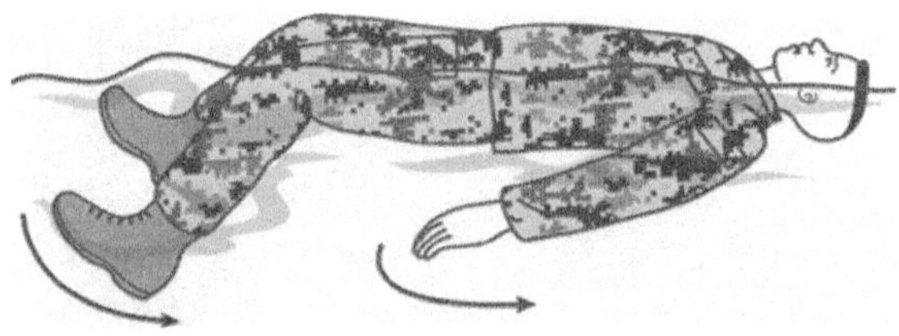

Near the finish of the pull, flex your knees to a 90 degree angle. The arm pull counteracts the resistance created by the knees.

Kick out your legs, and squeeze them together as the arm pull is completed (catch, power).

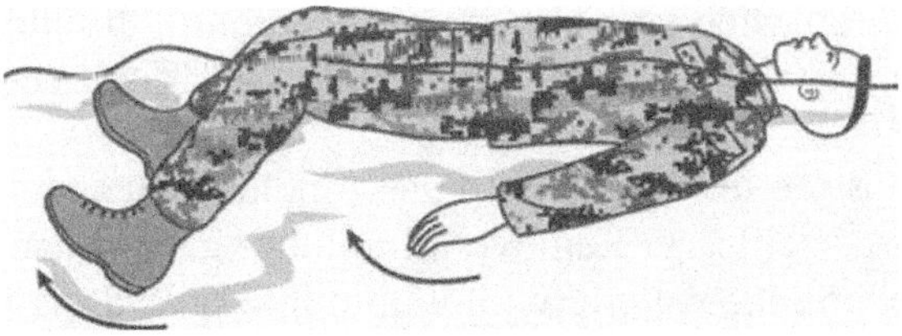

Glide through the water for 1 to 3 seconds or until your forward momentum decreases, then begin the next stroke as your momentum slows.

Chapter 2
Water Rescues

A drowning victim can panic and react with unexpected violence and can seize and inadvertently drown a rescuer. Therefore, if possible, a water rescue should be executed from a distance. Reaching, wading, or throwing methods are used in a distance rescue.

If a victim is too far away to use these methods, a swimming rescue may be necessary. Remove all combat gear, when possible, before entering the water. Swim within 2 to 6 yards of the victim to maintain a margin of safety, this allows you to reassess the situation and reassure the victim. If the victim is unconscious, use the wrist tow method or cross-chest carry method to pull the person to safety. If the victim is struggling, use a rear approach and then execute either a single armpit level off or a double armpit level off before towing the victim to safety.

If the victim does begin to overpower you, there are techniques that allow you to defend yourself without having to abandon the rescue. These techniques include the block, the wrist-grip escape, the front head-hold escape, and the rear head-hold escape. These techniques allow you to separate yourself from the victim, reassess the situation, and then attempt the rescue again.

> *Note:* In this chapter's illustrations, the rescuer is shown without a helmet.

Reaching Rescue Techniques

Reach

From a safe position at the water's edge, reach out to the victim. Talk constantly to calm the victim. Retain partial contact with land or some solid support structure (e.g., pier, bridge). If the victim is close but still beyond reach, extend an object (e.g., stick; pack; rifle with magazine removed, chamber empty, and muzzle pointing toward victim) that the victim can grasp. Pull the victim slowly to safety. Once the victim is close to shore, remove the victim from the water.

Note: You can also extend a foot to the victim if you can retain a secure grip on a solid support structure.

Reach From a Deck

Reach from a deck rescue can be executed by either a swimmer or a nonswimmer, and it can be used on an active or a passive victim. Be sure to reassure the victim during the rescue. To execute reach from a deck—

Lie prone on the deck with your body firmly anchored. To anchor your body, lie flat, spread legs apart, and extend one arm behind you with your palm down and on the deck.

Keep as much of your weight on the deck as possible and extend your free hand to the victim.

Grasp the victim's wrist from above, your thumb and index finger are facing you. The victim should never be allowed to grab you and put your life at risk. Therefore, never reach across the victim to grab his wrist, always grab the wrist that is the closest to you.

Keep your arm straight and locked out and pull the victim to the side of the deck. The victim should never be allowed to grab you and put your life at risk. Therefore, never pull the victim into you, always pull the victim into the side.

Arm Extension

An arm extension rescue can be used if you cannot reach the victim using the reach from a deck rescue technique and you must enter the water. An arm extension rescue technique is used for a victim who is either active or passive. Once you determine that a reach from a deck rescue technique is not viable—

Reassure the victim and quickly ease into the water while holding onto the deck with one hand.

Grasp the victim's wrist from above, your thumb and index finger are facing you. The victim should never be allowed to grab you and put your life at risk. Therefore, never reach across the victim to grab his wrist, always grab the wrist that is the closest to you.

Keep your arm straight and locked out and pull the victim to your side. The victim should never be allowed to grab you and put your life at risk. Therefore, never pull the victim into you, always pull the victim into the side.

Leg Extension

If the victim is beyond the reach of your arm, ensure that you have a firm grip on the deck, extend a leg to the victim and allow him to grab it, slowly bring the victim in closer until you can grab the victim's wrist that is holding onto your leg. Once you have a firm grasp on the victim's wrist, use the steps in the arm extension to pull the victim to safety. The leg extension rescue can only be used to rescue an active victim because the victim must be able to grab the rescuer's leg.

Wading Assist

Do not wade into water that is deeper than your chest. Talk constantly to calm the victim. If possible, do not touch the victim directly. Extend an object (e.g., stick; pack; rifle with magazine removed, chamber empty, and muzzle pointing toward victim) that the victim can grasp. Once the victim grasps the object, pull the victim slowly to safety.

Throw

If the victim is not within reach, use an expedient line to throw a lifesaving device to the victim. A lifesaving device can be any weighted item that floats (e.g., a canteen that is one fourth full of water). The lifesaving device must be secured to the end of the rope so that the rope will feed out from its coil when tossed. Talk constantly to calm the victim. Once the victim grasps the line or the lifesaving device, pull the victim toward you at a steady pace that keeps the victim's head above the water's surface. DO NOT pull so strongly that you break the victim's grip on the line. To prepare and use an expedient line and lifesaving device—

Tie a bowline at one end of the rope. (The appendix illustrates various types of knots.)

Unfasten the lid of a canteen.

Place the bowline around the neck of the canteen.

Refasten the lid so the canteen hangs from the bowline loop.

Place one end of the rope under the ball of your forward foot to secure it, either tie a knot at the end of the rope or tie an object at the end of the rope to create a block. Stand with your weight on the end of the rope. DO NOT tie the rope around your ankle.

Coil 20 to 30 yards of the rope, and hold it in a nonthrowing hand.

Place the canteen in your throwing hand.

Use an underhand throw to pitch the canteen and rope a short distance over the victim's head. Keep your nonthrowing hand open so the coil can unfold freely. The rope should trail across the victim's outstretched arms.

Retrieve the rope if the throw is inaccurate or the victim fails to grasp it. Recoil the rope as it is retrieved.

Divide the coil and throw again.

Lifesaving Approaches

Properly approaching the victim is as important as any other aspect of a rescue. Knowing the appropriate and safest method needed to approach a victim is critical in ensuring your safety and the victim's survival. Determining the victim's physical state (distressed swimmer, victim that is active and drowning, or victim that is passive and drowning) is crucial and will determine which type of approach you will execute.

Front Surface Approach

The front surface approach is typically performed when the victim is passive. You must remember that approaching the victim from the front is dangerous because a distressed swimmer or victim that is active can lunge toward you.

Rear Approach

If the victim is active, you must remember that his response can change rapidly. Therefore, approaching him from the front could be extremely dangerous. The victim could easily grab you and take you under the water during a state of panic. Therefore, use a rear approach when possible. When approaching a victim from the rear, reassure the victim until you get to within 2 to 6 yards of the victim; at that point, stop talking to the victim in order to maintain an element of surprise.

Approach Strokes

Before approaching a victim, you must evaluate the victim:

- Is the victim a distressed swimmer?
- Is the victim active and drowning?
- Is the victim passive and drowning?

Select your approach stroke based on your evaluation. Remember, the safety of the rescuer is paramount. Do not endanger yourself in an attempt to reach a drowning victim.

Crawl Stroke Approach Stroke

The crawl stroke (free style) is the fastest approach stroke. This stroke is used when the victim is passive and/or unconscious in the water. That is, the victim is face down, submerged or near the surface of the water, is not breathing, or is not moving. If any of these conditions exist, there is an obvious need to reach the victim as quickly as possible.

Execution of the crawl stroke approach stroke is the same as the crawl stroke (see pages 1-40 through 1-44), except that your head remains above the water's surface to allow for free breathing and maintaining eye contact with the victim.

Maintaining eye contact is critical. If the victim becomes submerged, you will have a better chance of locating him if you know his last location. Your distance from the victim will determine if your head remains out of the water the entire time you are executing the stroke. If you are 54 yards or less from the victim, you should maintain eye contact by keeping your head raised until you reach the victim. If you must swim more than 54 yards to reach the victim, make eye contact with the victim, place your face back in the water for four or five strokes, then raise your head again to regain eye contact with the victim (repeat these steps until you reach the victim).

Body position
The position of your body is horizontal to diagonal and influenced by the position of the head. This allows you to maintain eye contact with the victim.

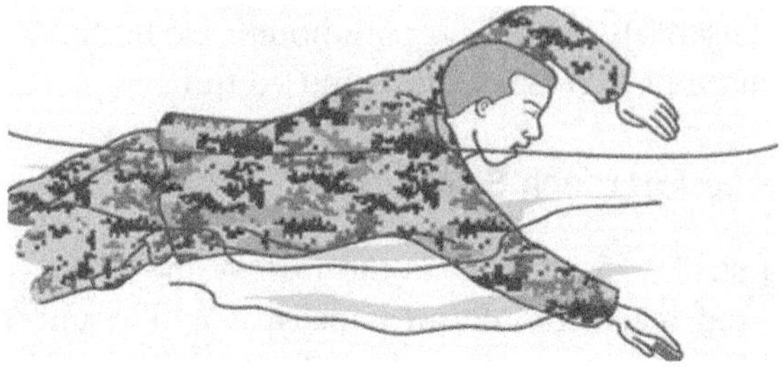

Arm action
While one arm is forward of your head to catch the water for the propulsion phase, extend your other arm alongside your body in what is known as the recovery phase. The recovery phase ends when your arm is extended forward of your head to begin the propulsion phase. In the propulsion phase, your arms are approximately shoulder-width apart and the pulling action is slightly wider and deeper to compensate for your raised head.

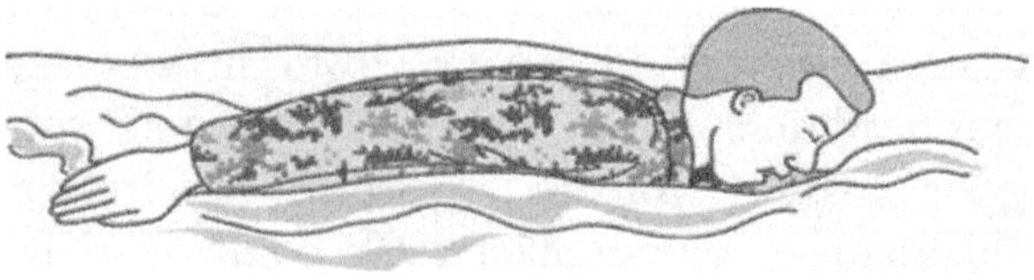

Leg action
Kick one leg in a downward motion while your other leg is recovering to the surface to prepare for the next kick (this is known as a flutter kick). This kick is used for propulsion and maintaining the lower body as horizontal to the water's surface as possible. When your head is raised, your knees are slightly bent in order to keep your feet near the surface.

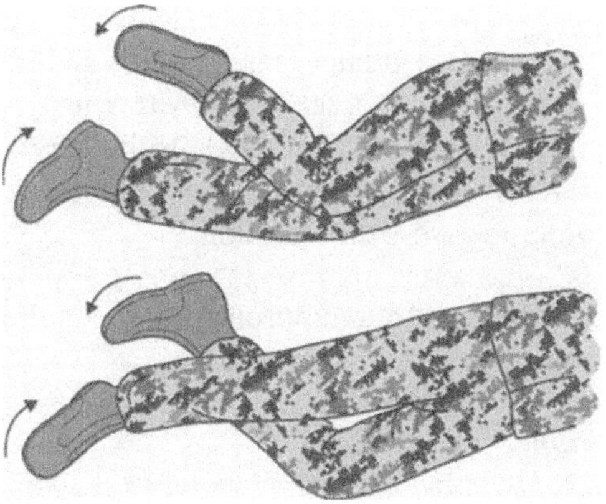

Breathing
If your face is above the water, breathe as needed (free breathing).

If traveling in excess of 54 yards to reach the victim, rotate your head to the right or left during the recovery phase and inhale when your face is out of the water or wait until your head is out of the water and facing forward while regaining sight of the victim.

Coordination
This stroke uses constant arm and leg action.

Breast Stroke Approach Stroke

The breast stroke approach stroke is used when a swimmer is distressed or a victim is active and drowning. This stroke is also used when the victim has a suspected spinal injury because it minimizes the wake created around the victim, minimizes the movement of the victim's head and neck, and helps prevent further injury.

If a victim is at a distance greater than 54 yards and if the swimmer is in distress or the victim is active and drowning in open water (ocean, lake), you should use the crawl stroke approach stroke to rapidly approach the victim, stopping approximately 11 yards from the victim in order to assess the situation.

Execution of the breast stroke approach stroke is the same as the breast stroke with a few minor deviations.

Body position
The position of your body is horizontal to diagonal and influenced by the position of the head. This allows you to maintain eye contact with the victim. It also allows you to communicate with the victim in order reassure him as you approach.

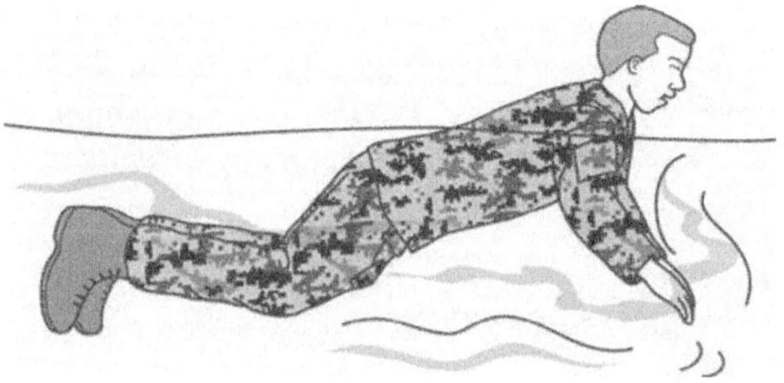

Arm action
Execute the arm action for the breast stroke, except that your arm pull is wider and deeper, which allows your head to remain above the water's surface.

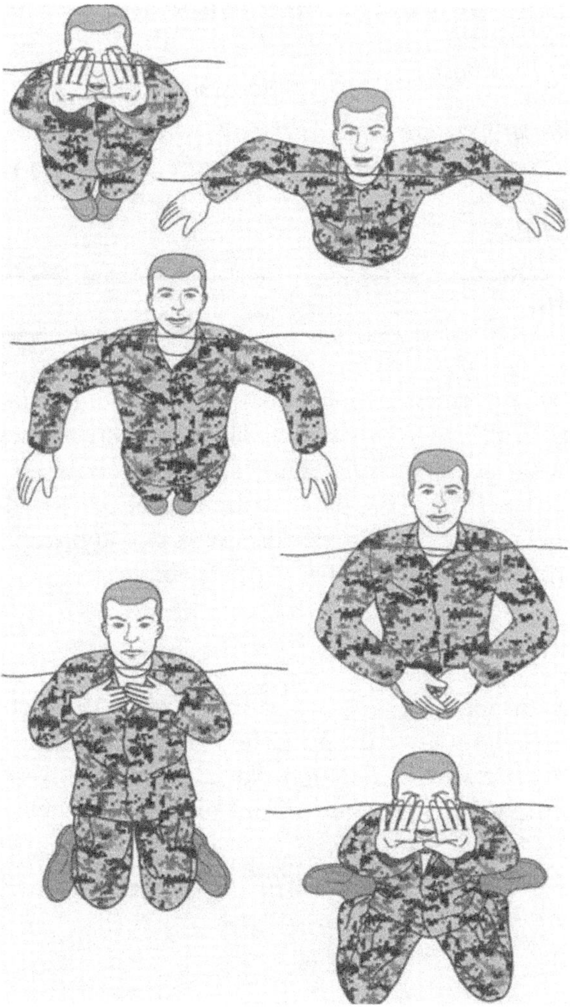

Leg action
Execute the leg action for the breast stroke, except that your legs open wider during the propulsion phase of the stroke.

Breathing
Keep your face above the water and breathe as needed (free breathing).

Coordination
Coordination is the same as the breast stroke, except that there is no glide period. You constantly stroke until you reach the victim.

Level Offs

A properly performed level off positions the victim's face up and horizontal to the water's surface. There are three types of level offs: front surface approach, single armpit level off, and the double armpit level off. The single armpit level off and the double armpit level off are done by performing a rear approach, which is the preferred approach when a victim is active.

Front Surface Approach

The front surface approach is performed when the victim is passive. Approaching a victim from the front may place you in danger because a distressed swimmer or a victim who is active may lunge toward you. To execute a front surface approach—

Stop 2 to 4 yards from the victim to reassess the situation and reassure the victim.

Determine which arm to use in order to rotate the victim into a face-up position (your right hand on the victim's right wrist or your left hand on the victim's left wrist).

Move into position and prepare to grasp the victim.

Turn sideways and move toward the front of victim.

Once in position, reach forward and grab the victim's wrist. Your thumb is on the inside of the victim's wrist, as if you were checking his pulse with your thumb. Your remaining fingers wrap around the victim's wrist.

Lean back immediately and execute a powerful scissor kick or inverted scissor kick and perform short, vigorous pulls with your free arm.

As the victim begins to move forward in the water, kick, pull, and twist outboard on the victim's wrist. The momentum created from the kick and the pulling and twisting action of the victim's wrist will rotate the victim into a face-up and horizontal position in the water.

Extend and lock out your towing arm down the length of your body and execute a wrist tow to move the victim to safety.

Single Armpit Level Off

Approach the victim.

Stop 2 to 6 yards from the victim to reassess the situation, reassure the victim, and maintain your margin of safety.

Approach the victim slowly and grab his armpit with your hand (your right hand to victim's right armpit or your left hand to victim's left armpit). Position yourself sideways to the victim and place your elbow in the center of his back.

Pull with your hand that is in the victim's armpit while pushing with your elbow that is in the victim's back to place the victim horizontal on the water's surface (face up). To assist in placing the victim horizontally, use your free arm to execute short, vigorous pulls and use your legs to execute a scissor kick or inverted scissor kick.

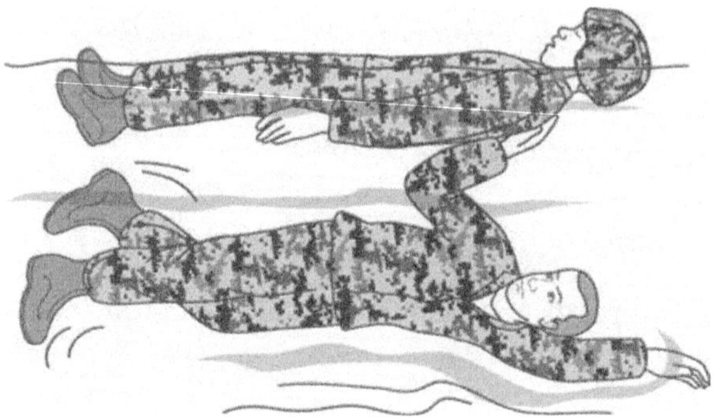

Once the victim is horizontal and on the water's surface, begin forward momentum by extending your towing arm to a fully locked-out position and executing a single armpit tow.

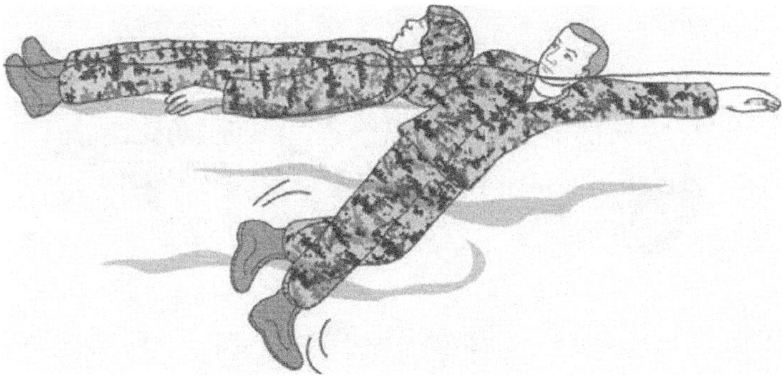

Double Armpit Level Off

Approach the victim.

Stop 2 to 6 yards from the victim to reassess the situation, reassure the victim, and maintain your margin of safety.

Approach the victim slowly and grab his armpits with both your hands. Place your elbows on his back.

Pull with your hands that are in the victim's armpits while pushing with your

elbows that are in the victim's back and use an inverted breast stroke kick to place the victim horizontal on the water's surface (face up).

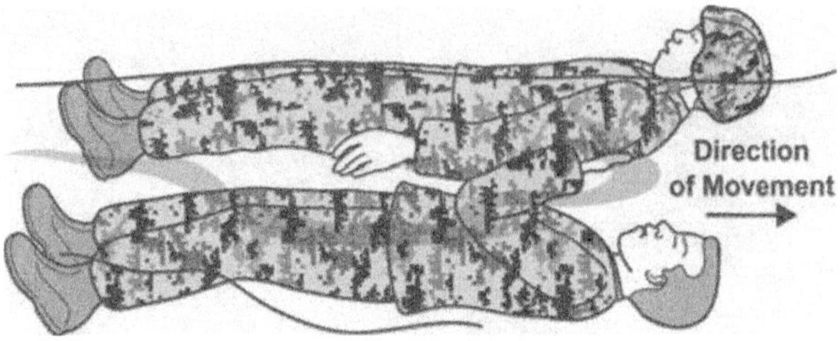

Once the victim is horizontal and on the water's surface, begin rearward momentum by slowly extending your arms into a fully locked-out position and executing a double armpit tow and inverted breast stroke kick.

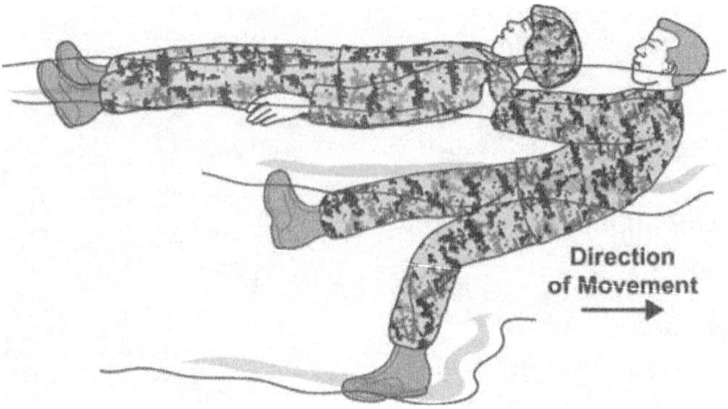

Rescue Techniques

Wrist Tow

Use the wrist tow method to rescue a victim who is floating face down. DO NOT use the wrist tow on a struggling victim. If time allows, remove your helmet and gear before attempting the rescue. Swim toward the victim using a modified breast stroke. Swim within 2 to 6 yards of the victim to maintain a margin of safety, this allows you to reassess the situation and reassure the victim. The following steps show proper front surface approach, wrist tow procedures:

Approach the victim from the front and grasp the underside of the victim's left wrist with your left hand or the right wrist with your right hand. Ensure that your thumb is on the underside of victim's wrist.

Lean back, pulling and kicking strongly to move the victim into a horizontal position.

Twist the victim's wrist to rotate the victim into a face-up position.

Swim toward safety using the lifesaving stroke.

> NOTE: The lifesaving stroke is a modified side stroke wherein the top arm is used to tow or carry a victim to safety and use a scissor kick or an inverted scissor kick is used for propulsion.

Keep a firm grip on the victim's wrist.

Keep your towing arm fully extended and along your side. This keeps the victim in the water column and prevents drag.

Ensure the victim's head does not go under water during the recovery.

Single Armpit Tow

You perform the single armpit tow after a victim has been properly leveled off. The single armpit tow uses the lifesaving stroke. To execute the single armpit tow—

Place one hand (top arm) in the victim's armpit. Your towing arm is straight and along your side.

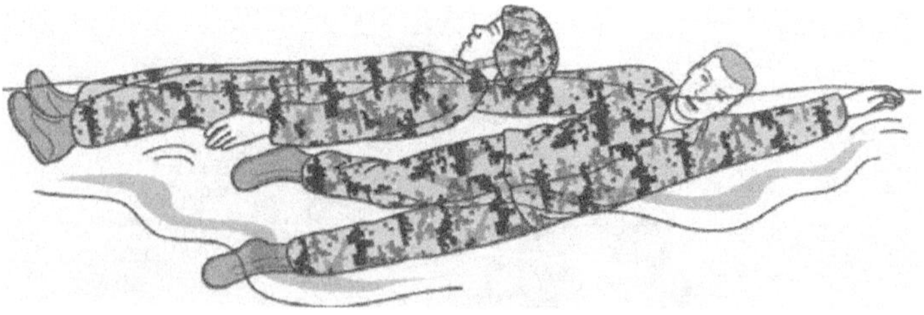

Use your lead arm (bottom arm) to execute short, vigorous pulls.

Execute a scissor or inverted scissor kick in a continuous and vigorous manner.

Use either free breathing or explosive breathing during the tow. During free breathing, keep your head above the water's surface and continuously reassure the victim. During

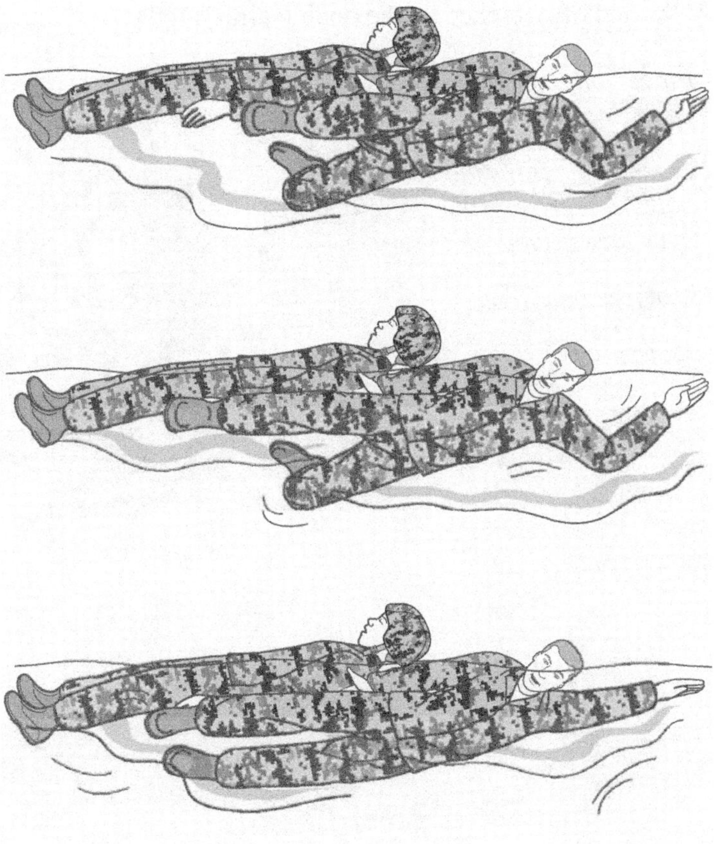

explosive breathing, put your head under the water to plane off your body's angle and reduce drag. Execute approximately two strokes, then lift your head up to breath and reassure the victim. Return your head back underwater for the next few strokes.

Double Armpit Tow

You perform the double armpit tow after a victim has been properly leveled off. To execute the double armpit tow—

Place both hands in the victim's armpits.

Extend both arms fully along your body. You are on your back, and your face is clear of the water.

Use free breathing.

Reassure the victim at all times.

Use the inverted breast stroke kick (it is the only kick that can be used due to your body position) to tow the victim to safety. (The inverted breast stroke kick is the same kick used in the elementary backstroke.) Your kick must be continuous and vigorous in order to keep the victim's face above the water.

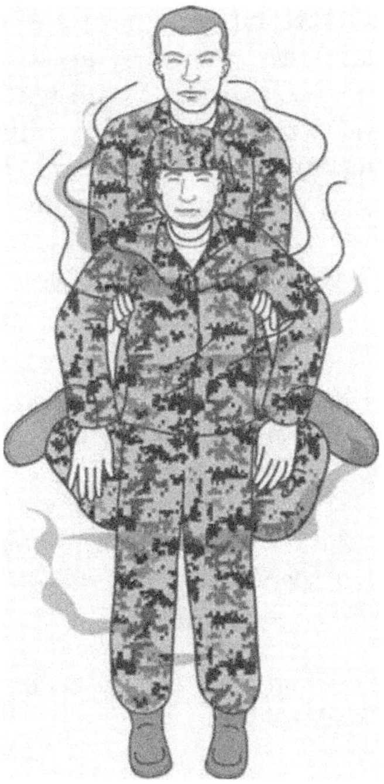

Collar Tow

You perform the collar tow after a victim has been properly leveled off. To perform a collar tow using the victim's blouse—

Maintain control of the victim by grasping his armpit with one hand and then with your free hand grasp either his combat gear or his blouse between his shoulder blades. If grasping the blouse, grasp the material with your palm up, then turn your hand over to tighten the material.

Release the victim's armpit once control is established and execute the lifesaving stroke to tow the him to safety.

Cross-Chest Carry

Use the cross-chest carry to carry a victim to safety if the victim is struggling or when moving through heavy surf. Remove your helmet and gear before attempting a rescue. Talk constantly to calm the victim. Swim toward the victim using a breast stroke approach stroke. Swim within 2 to 6 yards of the victim to maintain a margin of safety, this allows you to reassess the situation and reassure the victim. The following steps illustrate proper cross-chest carry procedures:

> **CAUTION:** The cross-chest carry causes fatigue even if you are in excellent physical condition.

Use a level-off technique to place the victim in a horizontal, face-up position.

Retain a grip on the victim with one hand. Reach over the victim with your free hand to encircle the victim's chest. Place your free hand on the victim's opposite rib cage, just below his armpit.

Release your grip once you have a secure hold on the victim's chest.

Swim toward safety using the lifesaving stroke while keeping a firm grip on the victim's chest and your hip on his back.

This procedure brings the victim's face and shoulders clear of the water, and typically the victim stops struggling. Sometimes, however, the victim will struggle during the swim to safety. If this happens, either tighten your grip on the victim or defend yourself with one of the techniques discussed on pages 2-25 through 2-30.

Tired Swimmer's Assist

You must maintain a 2 to 6 yards margin of safety from the victim at all times while you are getting into position to perform the tired swimmer's assist. Once behind the victim, you extend one arm, hand straight. Place your straight hand underneath the victim's opposite armpit. Maintain a 45 degree angle away from the victim, your arm is locked out. Assist the victim in propulsion until both of you reach safety.

Defense Against a Drowning Victim

DO NOT sacrifice your life in an attempt to save the victim. A struggling drowning victim poses great danger to anyone nearby. Driven by panic, the victim can grab you with great strength in an effort to climb out of the water. This can result in death for both you and the victim. The following techniques can be used to defend against a drowning victim, but the best defense against attack by the victim is to stay out of reach.

Block

The block prevents the victim from grabbing you if you have approached from the front. If the victim lunges toward you, react as follows:

> *Place one or both open hands against the victim's upper chest; being careful to avoid the victim's face, neck, and abdomen.*

Lean backwards and submerge rapidly. Keep your blocking arm(s) extended.

Swim underwater and away from the victim, and return quickly to the surface.

Stop 2 to 6 yards from the victim to reassess the situation.

Determine an appropriate course of action.

Wrist-Grip Escape/Wrist-Grip Escape Alternative

The wrist-grip escape is used when a victim grabs your arm or wrist. Submerge the victim quickly by reaching across with your free hand, pushing down on the victim's shoulder to submerge him, and kicking to propel yourself upward. While keeping the victim submerged with your hand on his shoulder, give three hard jerking pulls with your trapped hand in an attempt to break free from his grasp. Once free, swim clear of the victim and reassess the victim's condition.

If the wrist-grip escape is unsuccessful in freeing your hand, use the wrist-grip escape alternative: use your free hand to grab your trapped fist, rotate thumbs up, apply bone-on-bone contact with the victim's arm, pry your hand out of his grasp, and quickly swim away from the victim.

Front Head-Hold Escape

The front head-hold escape allows you to escape when you are facing a victim who is gripping you around your head and neck. To execute the front-head hold escape—

Take a quick breath and tuck your chin into your shoulder to protect your throat.

Clap your hands above your head (three times) to submerge instantly. This drags the victim below the water, lifts his arms from around your neck, and, typically, he releases his grasp in order to get back to the surface. If he doesn't release his grasp, apply pressure to the victim's brachial pressure points (which are located inside of the upper arm, above the elbow).

Thrust the victim's arms up and away.

Keep your chin tucked to protect your throat, and swim underwater away from the victim and return quickly to the surface at the ready position.

Stop 2 to 6 yards from the victim to reassess the situation.

Determine an appropriate course of action.

Rear Head-Hold Escape

The rear head-hold escape allows you to escape when a victim is gripping your head and neck from the rear. To execute the rear head-hold escape—

Take a quick breath. and tuck your chin down, turn your head to either side, and raise your shoulders to protect your throat.

Take a strong stroke, clap your hands above your head (three times), and submerge instantly. This drags the victim below the water and, typically, he releases his grasp in order to get back to the surface. If he doesn't release his grasp, apply pressure to the victim's brachial pressure points (which are located inside of the upper arm, above the elbow).

Thrust the victim's arms up and away.

Twist your head and shoulders until free.

Swim underwater away from the victim and return quickly to the surface.

Stop 2 to 6 yards from the victim to assess the situation.

Determine an appropriate course of action.

Administering First Aid/Rescue Breathing

To administer first aid/rescue breathing, remove the unconscious victim from the water when possible and stop any bleeding.

First, open the airway and check for breathing. If the victim is not breathing, give two breaths of air. Check for a pulse. If there is a pulse but the victim is not breathing, continue rescue breathing. If there is no pulse, start cardiopulmonary resuscitation (CPR). For detailed information on administering CPR and rescue breathing, refer to American Red Cross CPR skill cards.

WARNING

> If the victim has no pulse and is not breathing, administer CPR immediately. If the victim does have a pulse but is not breathing, give rescue breathing only. If the victim has a pulse and is breathing, DO NOT give CPR—CPR could prove fatal.

Next, protect any wounds from exposure, and treat for shock.

Chapter 3
Treatment of Casualties and Avoidance of Dangerous Marine Life

To survive in the water, you will face many challenges other than just being able to swim. For example, injuries and/or fatigue can lead to drowning and exposure to the elements can lead to hypothermia or heat injuries and may require medical attention. Finally, your presence in the water may be seen as a threat to or food source for a variety of marine life and you may be attacked. This chapter advises you on how to minimize these risks in order to stay alive.

Drowning

Drowning is suffocation by a liquid. A drowning victim inhales water into the lungs or the throat closes by reflex so that little or no water can enter the windpipe. In either case, a victim can no longer breathe.

Symptoms

One symptom of drowning is that the victim may call for help and has an expression of dread or panic. But typically a victim that is active and drowning may not call for help because he is trying to conserve his air and will not speak.

Another symptom of drowning is when the victim thrashes at the water's surface. If the victim stops or grows calmer, he has likely been overcome by fatigue, hypothermia, or a lack of air. At this stage, the victim usually has 20 to 60 seconds before going under the water's surface.

Treatment

If the victim is not breathing, begin rescue breathing. Place the victim on his back, tilt head back to open airway, pinch the nose, and give two full breaths. If the victim does not inhale during the first two breaths, reposition his head and attempt two more breaths. Check for a pulse. If a pulse is present, but the victim is still not breathing, continue rescue breathing. If a pulse is not present, begin CPR. See MCRP 3-02G, *First Aid*, for rescue breathing and CPR details.

WARNING

If the victim has no pulse and is not breathing, administer CPR immediately. If the victim does have a pulse but is not breathing, give rescue breathing only. If the victim has a pulse and is breathing, DO NOT give CPR—CPR could prove fatal. Continue first aid until medical help arrives.

A victim who is not breathing and has no pulse may appear dead. DO NOT decide that death has occurred. Continue with the prescribed treatment. A corpsman or medical officer should decide whether the victim can or cannot be revived.

Hypothermia

Hypothermia is the abnormal lowering of the body's internal (or core) temperature to 95 degrees or below. It occurs when the body loses heat faster than it produces heat. A hypothermia victim loses the ability to move quickly, becomes mentally sluggish, slips into semiconsciousness, lapses into a coma, and dies when internal body temperatures drop too low.

The chilling effects of cold air, wind, or water can produce hypothermia. Water poses the greatest threat because it transfers heat 25 times faster than air. Depending on the water's temperature, a victim can succumb to hypothermia within a few minutes. The body's sudden contact with cold water can also set off a body reaction known as the mammalian diving reflex. This reflex can greatly increase survival time (especially for women and children) in or under cold water. The mammalian diving reflex shuts off blood circulation, except for the flow between the heart, lungs, and brain. The small amount of oxygen left in the blood and lungs is saved for the body's vital organs. This reflex has allowed people to survive being under cold water for an extended period of time. Therefore, a cold water drowning victim should be treated as if still alive even though the victim is not breathing, has no pulse, and may appear dead. DO NOT decide that death has occurred. Continue with the prescribed treatment. Victims of hypothermia can appear to be dead when they are not. A corpsman or medical officer should decide whether the victim can or cannot be revived.

WARNING

If the victim has no pulse and is not breathing, administer CPR immediately. If the victim does have a pulse but is not breathing, give rescue breathing only. If the victim has a pulse and is breathing, DO NOT give CPR—CPR could prove fatal. Continue first aid until medical help arrives. Check for a pulse for at least 45 seconds.

Symptoms

Once the body's core temperature drops, the victim will show one or more of the following symptoms:

- Violent and uncontrollable shivering as the body tries to warm itself.
- Slow or slurred speech.
- Disorientation or poor coordination.
- Loss of skin color.
- Blue and pinched lips.
- A slowing or stopping of shivering that progresses into a rigid torso and limbs.

Survival Time

A hypothermia victim's survival depends on the water's temperature and the time spent in the water. A small body build cools faster than a large build. Children cool faster than adults.

To increase your chance of survival in the water, utilize the HELP position described on page 1-37. Extra clothing and inactivity (remaining motionless in the water) can also increase your survival time.

Treatment

A hypothermia victim must be warmed to prevent further heat loss; therefore, treatment should begin as soon as possible. Consciousness of the victim determines the treatment that should be pursued. See MCRP 3-02G for specific treatment information. The following treatment procedures are recommended:

- If the victim is conscious, give the victim warm fluids. Give candy or sweetened foods to a victim who is able to eat.
- If the victim is unconscious, place him on his back with his head tilted back to ensure an open airway.
- DO NOT massage the victim. Massage can break blood vessels and create swelling, internal pressure, and blocked blood circulation.
- DO NOT give alcohol to the victim. Alcohol lowers the victim's body temperature.
- Shock is a possibility, treat accordingly.
- Seek medical help immediately.

If you are able to remove the victim from the water, apply the following steps when possible:

- Get the victim into shelter.
- Remove the victim's wet clothing.
- Put the victim in dry clothing.
- Place the victim in a sleeping bag if one is available. It may be necessary to place another Marine in the sleeping bag with the victim.
- Place as much insulation as possible between the victim and the ground.
- Use hot water bottles, electric blankets, or blankets heated in an oven or by a campfire to warm the victim's neck, groin, and the sides of the chest.

CAUTION: DO NOT apply heat to extremities.

Heat-Related Injuries

Heat-related injuries include heat cramps, heat exhaustion, and/or heat stroke. They can occur when a Marine—

- Is exposed to extreme heat, such as from the sun or a combination of high air temperatures and water temperatures.
- Does not wear proper clothing or gets overexposed to the sun.
- Becomes dehydrated.

With heat cramps, muscles may cramp in the arms, legs, and/or stomach. The victim may sweat excessively. To treat, create improvised shade for the victim and have him drink water.

With heat exhaustion, the victim sweats heavily; presents pale, moist, cool skin; and complains of a headache, weakness, dizziness, and/or loss of appetite. A victim may also experience heat cramps, nausea, vomiting, an urge to defecate, chills, rapid breathing, confusion, and a tingling sensation in the hands or feet. To treat, pour water over the victim and fan him to speed up the coolant effect of evaporation, have the victim drink freshwater, and attempt to provide shade.

With heat stroke, the victim stops sweating, which results in red, flushed, hot, dry skin. The victim may first experience headaches, dizziness, nausea, fast pulse and respiration, and/or seizures and mental confusion. The victim can become unconscious and die if not treated quickly. To treat, create improvised shade for the victim. Have him drink water and elevate both legs if possible.

Burns

Apply a field dressing (or the cleanest material available) to the burn. Give sips of water to a casualty who is conscious and not nauseated. When treating burn casualties—

- DO NOT remove clothing stuck to the burns.
- DO NOT break any blisters.
- DO NOT apply grease or ointments to the burns.

For electrical burns, check for both an entry and exit burn from the passage of electricity through the body. An exit burn may appear on any area of the body, not necessarily opposite the entry burn.

For burns caused by wet or dry chemicals, flush the burns with large amounts of water and cover with a dry dressing.

For burns caused by white phosphorus, flush the area with water, then cover with a wet material, dressing, or mud to exclude air and keep the white phosphorus particles from burning.

For laser burns, apply a field dressing.

Common Medical Problems Associated with Sea Survival

Seasickness

Seasickness is the nausea and vomiting caused by the bobbing motion created by the wave action of a flotation device. Seasickness can result in—

- Dehydration and exhaustion.
- A loss of the will to survive.
- Others becoming seasick.
- Unclean conditions.

To treat seasickness—

- Wash both the Marine and the flotation device to remove the sight and odor of vomit.
- Keep the Marine from eating food until the nausea is gone.
- Have the Marine lie down and rest.
- Give the Marine seasickness pills if available. If the Marine is unable to take them orally, the pills should be inserted rectally for absorption by the body.

Saltwater Sores

Saltwater sores occur when skin that has abrasions or is cut is exposed to saltwater. The sores may form scabs and pus. Do not open or drain. Flush the sores with freshwater, if available, and allow to dry. Apply antiseptic, if available.

Blindness/Headache

Irritants or the effects of the sun's rays reflecting off the water can cause temporary blindness or headaches. If flames, smoke, or other irritants get in your eyes, flush the eyes immediately with saltwater, then with freshwater, if available. Apply an ointment, if available. Bandage both eyes for 18 to 24 hours or longer if the damage is severe. If glare from the sky and water causes your eyes to become bloodshot and inflamed, bandage the eyes lightly. Try to prevent this problem by wearing sunglasses or goggles with a sunglass insert.

Constipation

This condition is a common problem associated with dehydration. For constipation, do not take a laxative if it is available; this causes further dehydration. Drink freshwater, if available.

Sunburn and Dehydration

The sun's rays reflect at all angles off the waves of the water; therefore, sunburn and dehydration are serious problems in sea survival. Try to prevent sunburn by—

- Erecting an improvised canopy, with available floating materials, to provide shade.
- Wearing your soft cover or using a cloth, such as a handkerchief, to cover your head.
- Covering your skin with sunscreen or lip balm from your first aid kit. Your lips, nostrils, eyelids, the backs of your ears, and the skin under your chin sunburns easily. If enough sunscreen cream is available, all exposed skin should be covered.

Dehydration is caused by the loss of the body's vital fluids. Dehydration in saltwater may result from a combination of factors such as a lack of water, the effects of saltwater on skin tissue, sunburn or vomiting from seasickness, and other causes. Sleep and rest and reduced water and food intake are the best ways of enduring periods of exposure. The following measures will delay the effects of dehydration:

- DO NOT drink saltwater.
- DO NOT drink urine.
- DO NOT drink alcohol.
- DO NOT smoke.
- DO NOT EAT unless water is available.

Dangerous Marine Life

You may see many types of marine life around you, and some are more dangerous than others. Generally, sharks are your greatest danger, followed by barracudas. However, most marine life will not deliberately attack a human. The most common injuries from marine life are wounds from bites, stings, or punctures. With the exception of sharks and barracudas, most injuries are a result of either trying to catch game or from contact abrasion with marine life. To treat an injury that results from a dangerous marine animal bite—

- Pack the wound with gauze and then apply a pressure bandage, if available.
- Treat for shock.
- Obtain medical attention as soon as possible.

For a more detailed information on the treatment of casualties and dangerous marine life, see MCRP 3-02F and MCRP 3-02G.

Sharks and Barracudas

Only about 20 percent of all shark species are known to attack people. Sharks have an acute sense of smell, and the smell of blood in the water will draw them to their prey. They are also very sensitive to any abnormal vibrations in the water; therefore, the sound caused by a struggling swimmer or underwater explosions will attract them.

Barracudas are bold and inquisitive fish. They have been known to attack men who are wearing shiny objects. Barracudas may charge at lights or shiny objects at night.

A group of swimmers can maintain a 360 degree watch while in the water. Therefore, to protect yourself from sharks and barracudas, stay with other swimmers. The group members can either frighten away or fight off sharks or barracudas better than one person can. Keep all your clothing on, including footwear. Historically, sharks attack unclothed individuals in groups first, mainly in the feet. Clothing also protects you against abrasions from a shark's tough skin should the shark brush up against you. Avoid urinating heavily, let urine dissipate between discharges. If you must defecate, do so in small amounts and toss it as far away from the group as possible. Do the same if you must vomit.

If attacked, the use of firearms by swim sentries should be used with extreme caution because of the risk of injury to other swimmers. If unarmed or unable to make an improvised weapon, kick and strike the shark. Avoid using your bare hands to strike the shark, injury can result to your hands due to a shark's tough skin. Target areas on a shark are the gills, eyes, and underbelly. Blows to the snout are also not recommended because a shark will tilt its head up and thrust its jaws forward when biting.

Sea Snakes

Sea snakes are venomous and sometimes found in mid-ocean. They are unlikely to bite unless provoked. Your best protective measure is to avoid them.

Poisonous Fish

Many reef fish are toxic, and can kill you if eaten. Generally, poisons are present in all parts of the fish, but especially concentrated in the liver, intestines, and eggs.

Turtles and Moray Eels

Turtles and moray eels normally inflict minor bite wounds. Treat this type of injury by cleaning the wound.

Corals

Coral, dead or alive, can inflict painful cuts. Clean all coral cuts thoroughly. DO NOT use iodine to disinfect any coral cuts. Some coral polyps feed on iodine and may grow inside the flesh.

Jellyfish, Portuguese Man-of-War, Anemones, and Others

This group of marine animals inflicts injury by stinging their victims with their tentacles. Contact with their tentacles produces burning pain, a rash, and small hemorrhages on the skin. Shock, muscular cramping, nausea, vomiting, and respiratory distress may also occur. Gently remove the clinging tentacles with a towel and wash or treat the area. Use diluted ammonia or alcohol and talcum powder to treat the injury if available.

Spiny Fish, Urchins, Stingrays, and Cone Shells

These animals inject their venom by puncturing the skin with their spines. General signs and symptoms include swelling, nausea, vomiting, generalized cramps, diarrhea, muscular paralysis, and shock. Deaths are rare. Treatment consists of soaking wounds in hot water, if available, to deactivate heat-sensitive toxins.

Chapter 4
Negotiating Water Obstacles

Marines face water obstacles in saltwater, freshwater, and brackish water (where freshwater and saltwater meet). These water environments differ considerably and pose distinct problems for Marine tactical units and swimmers. Saltwater obstacles include tides, surfs, and currents. Brackish water obstacles include back bays. Freshwater obstacles include rivers and canals.

Tides

Tides are periodic changes in the surface levels of oceans, bays, gulfs, inlets, and rivers. The Moon's and Sun's gravitational pulls cause tides. Depending on the situation, tides can either help or hinder Marines in their endeavors to conduct amphibious or riverine operations. Tides can provide sufficient depths to allow for the passage of landing craft or boats over reefs, trees, and other underwater obstructions. On the other hand, tides can render a river fording site unusable to tactical vehicles. Direction, level of change, and amount of change determine tidal nomenclature.

Tides that show change in direction are flood tides and ebb tides. Rising tides are known as flood tides. Falling tides are known as ebb tides.

Tides that show extreme levels of change are high tides and low tides. High tide is the period when water is at its greatest depth. Low tide is the period when water is at its most shallow depth.

Tides that show the least and most amount of change are neap tides and spring tides, respectively. Neap tides have the least amount of change in water levels between high and low tide. Neap tides occur at the half moon, when the Sun and the Moon are aligned at a 90-degree angle with the Earth. In this position, the Sun's and Moon's gravitational pulls offset each other. Spring tides have the highest floods and lowest ebbs. Spring tides occur at or shortly after the new moon or full moon when the Sun, Moon, and Earth are approximately in line. In this position, the Sun's and Moon's gravitational pulls are combined.

Surf

Waves break upon entering shallow water and create surf. The offshore area where waves break is the surf zone, which can present many hazards. Breaking waves often trap air bubbles and create a foamy appearance. Bubbles lower the water's density and decrease buoyancy. Move through foamy surf as quickly as possible. The type of wave determines your survival technique.

Wave action moves you toward shore. Lie on your back or side with your head pointing in the direction of the beach and your feet pointing into the waves.

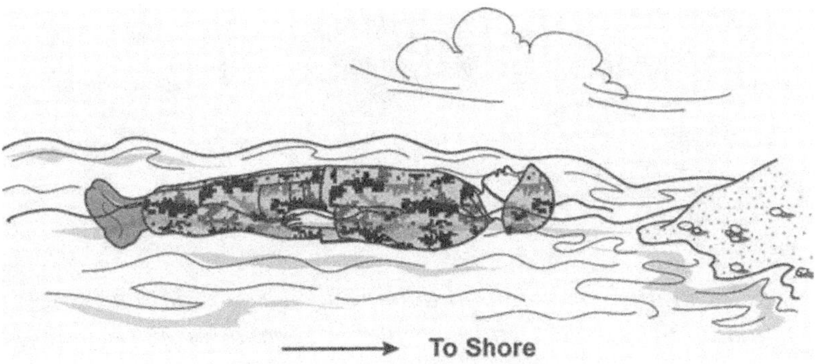

As one wave approaches the beach, another drains away; relax and do not swim against the draining water. When a new wave is within about 3 yards, start swimming toward shore. Continue to swim until the wave lifts you and moves you toward the beach. Once the wave loses forward momentum, relax and repeat the cycle. If nearing rocks, turn your body and approach feet first to reduce the chance of striking your head and arms.

Plunging Waves

A plunging wave is a breaker whose crest curls forward and falls ahead of its base. Because of its power and underwater turbulence, a plunging wave poses the greatest surf threat. If caught in a plunging wave, you can be pulled underwater and pitched about violently. This can cause you to panic, which can increase your chance of drowning. Perform the following steps to escape a plunging wave:

Tuck into a ball with your head against your knees and your forearms locked around your legs, just below the knees.

Relax in this position until the turbulence subsides and you float to the surface. This can take 30 seconds or more.

Swim toward shore.

> *Note:* If threatened by another plunging wave, dive underwater into the wave.

Spilling Waves

A spilling wave does not break. Instead, its crest slides forward without curling. A spilling wave creates less turbulence and poses less of a threat than a plunging wave. If caught in a spilling wave, relax and let the wave carry you to shore.

Surging Waves

A surging wave occurs on a beach with a steep underwater gradient. It never really breaks, but the crest rises while the base slides up the beach with great force and speed. Once the wave reaches its highest point on the beach, it rushes back as quickly as it surged forward. If you are standing on the bottom when a surging wave advances or retreats, the wave can knock you off your feet and pull you into the surf zone. Do not try to stand or walk on the bottom. Swim toward the beach as soon as possible.

Currents

Offshore Currents

An offshore current occurs outside the surf zone. Typically, it occurs at bay entrances, in island channels, and between islands and the mainland. An offshore current flows parallel to or away from shore. If the offshore current is created by tides, its current strength and direction vary at different times of the day.

If caught in an offshore current, you may be carried in a direction you do not want to go. DO NOT try to swim directly to safety. If the current is moving directly away from the shore, relax and wait until the current dies out or turns toward land. Once the current subsides, use any survival stroke to swim toward shore. If the current is moving parallel to shore, use any stroke to move at an angle across the current and toward shore.

Rip Currents

A rip current occurs when waves pile water against the shore faster than the water can drain. The water flows rapidly along the beach until it is deflected seaward by a bottom obstruction. Then the rip current flows through the surf zone and into open water at a speed of up to 2 knots. This action can cut deep trenches in the sand. A rip current dies out once it reaches open water (usually within a few hundred yards of the shore).

A rip current can pull you out to the open sea. If caught in a rip current, DO NOT try to swim against the current. A rip current moves faster than most people swim, and it is impossible to swim to shore once caught in it. Relax and stay afloat until the current runs out. Once the current subsides, use a survival stroke to move parallel to the shore until you are out of the current, then begin swimming toward shore.

Littoral Currents

A littoral current occurs when a wave breaks against a beach at an angle. This current flows parallel to the shoreline and does not pose a great threat. If caught in a littoral current, use the combat travel stroke to swim across it at an oblique angle.

Back Bays

Once on the beach, you often face one or more rows of low hills called dunes. Behind the dunes, you may encounter a low-lying stretch of ground thickly covered with scrub trees and bushes. This area gives way to wetlands known as back bays. Back bays consist of muddy islands that are almost submerged during flood tides and separated by channels of brackish water of varying depths. Channel bottoms usually contain soft mud. Back bays pose major obstacles to vehicular traffic.

Infantry can cross back bays, but only with great effort. If crossing back bays by foot, consult detailed navigation charts and use the following guidelines to plan your route:

- Avoid water less than waist deep; walking in shallow water or soft mud is extremely tiring.
- Avoid back bay islands; these low-lying islands are usually too muddy to support foot traffic.
- Seek out deep water; floating with a pack is less tiring than walking through shallow water or soft mud.
- Seek out sand, shell, gravel, or stone bottoms; these firmer bottoms generally ease travel and help conserve energy.

Rivers and Canals

A river is a large, natural stream of water that empties into a larger body of water. The slope of the riverbed and the volume of water in the river determine its current.

Canals resemble small rivers or streams in their width and depth, but usually lack any significant current. Climbing out of these waterways can be difficult if the canal is flanked by steep banks.

Chapter 5
Fording Waterways

WARNING

Fords are dangerous. Cross as quickly as possible.

A ford is any site in a river, stream, or canal where the water is shallow enough for troops or vehicles to cross without using flotation devices. Canal bottoms are usually too soft to support fording vehicles, and wading infantry frequently stumble.

The tactical situation dictates the location of the fording site. Seek fords that are protected from enemy observation and that allow for adequate supporting fires. A night fording takes at least one and a half times as long as a daylight fording.

Note: In this chapter, the term "river" refers to rivers, streams, and canals.

Selection of a Ford Site

The following table identifies desirable fording characteristics:

Characteristics	Comments
Concealment	The ford hides personnel and vehicle movement from enemy observation.
Accessibility	The ford should have low banks with gentle gradients. This allows a free flow of traffic at both the entrance and the exit.

Characteristics (Continued)	Comments (Continued)
Slow Current	The ford's current should not exceed 1.5 meters per second if possible.
Firm Footing	The ford's bottom, entry, and exit composition should be firm enough to support traffic. Do not drive a vehicle over any bottom composition that a 2-inch diameter stick can be pressed into more than 1 or 2 inches.
Gently Sloped Channels	The ford's entry and exit points should be gently sloped. If possible, locate a portion of the stream where the channel is not actively shifting.
Depth	The fording depth is less than or equal to the least capable vehicle.

Determine the Slope

Units move into and out of water faster and more quietly if entry and exit points are not steep or muddy. Slope is the amount of change in ground horizontal distance (run) and in vertical elevation (rise) from one point to another. Slope is usually expressed as a percentage. You can use a clinometer, map, or line of sight and pace to measure the percentage of a slope.

Clinometer

A clinometer measures percentage of a slope. It is a component of the M9 armored combat excavator, and organic to many Marine engineer units. However, engineers are also taught (and often use) field-expedient methods to determine slopes and gradients.

Map

A map measures the horizontal distance along a desired path. Determine the difference in elevation between the path's starting and ending points. From a map's scale, you can determine the distance between two points. From a map's elevation lines, you can determine the difference in height between the same two points. Both figures must be in the same unit of measure (e.g., feet, meter). Divide the elevation (rise) by the distance (run) and multiply by 100.

Rise = 165 - 120 = 45 meters

Run = 200 meters

% Slope = $\frac{45}{200}$ **x 100 - 22.5%**

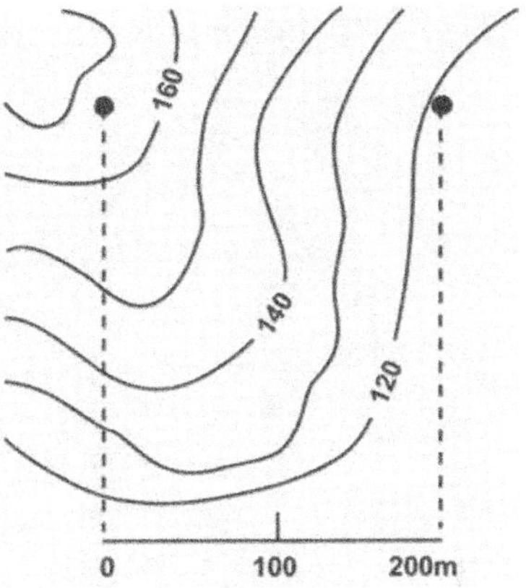

Line of Sight and Pace

To determine line of site and pace, stand at the bottom of the slope, keep your eyes level, pick a spot on the slope, then pace the distance. The number of paces multiplied by a standard measure of 0.75 meter determines the run. The eye-level height (usually 1.5 to 1.7 meters) determines the rise. Repeat this procedure until you have covered the entire distance you want to measure for each spot (vertical and horizontal). Add the vertical distances to provide total rise and the horizontal distances to provide total run.

Determine the Current Speed

Current speed increases as channels narrow. It may be necessary to locate a wider ford location to obtain a slower stream current. The following steps are used to calculate the speed of the current:

Determine points A and B along a channel. Then measure the distance between those two points.

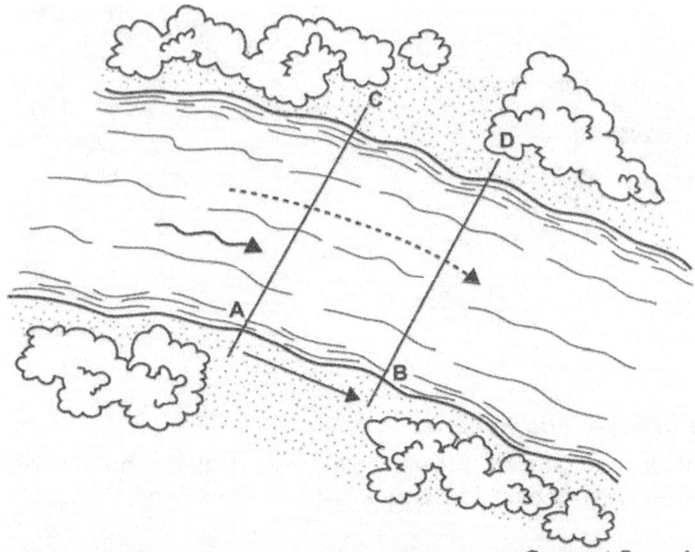

$$\frac{\text{Distance (meters) Between Points A and B}}{\text{Time (seconds) Between Points C and D}} = \text{Current Speed (meters per second)}$$

Sight directly across the water from points A and B to locate points C and D.

Throw a floating object (e.g., a stick) upstream from points A and C. Observe the object as it floats toward points B and D.

Subtract to find the time it takes for the object to float from start to finish.

Do not attempt to swim across currents that are moving faster than 1.5 meters per second. Equivalents of this speed include—

- Quick-time march rate of 120 counts per minute with one, 30-inch step at each count.
- 5 feet per second.
- 3.5 miles per hour.
- 5.5 kilometers per hour.

Measure River Width

A river's width can be estimated from the width of its symbol on a scaled topographic map. If this is not possible, use the following compass techniques:

Stand at the water line (A).

Shoot an azimuth to a point on the opposite bank (B).

Move upstream or downstream until you are at a point (C) where you can shoot an azimuth 45 degrees larger or smaller than the original azimuth.

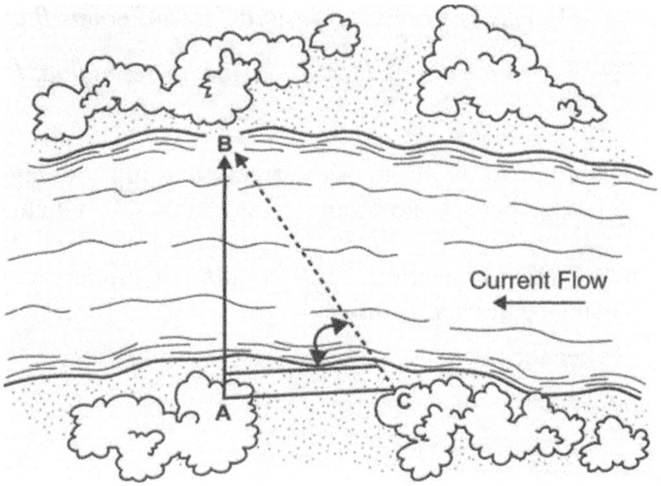

Measure the distance between points A and C. The distance calculated equals the river's width.

Calculate Downstream Drift

A river's current causes personnel and equipment to drift downstream. If personnel and equipment are aimed straight across the river, they will sideslip downstream as they move across the current to the other shore. Therefore, personnel and equipment crossings must compensate for the effects of a river's current and entries are usually made upstream of the desired exit point. Use the formula below to calculate downstream drift.

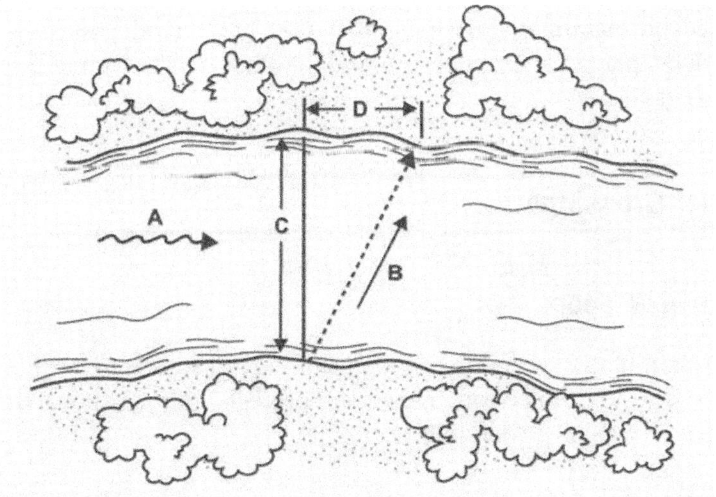

$$\frac{\text{Current Speed (A)}}{\text{Crossing Speed (B)}} \times \text{River Width (C)} = \text{Downstream Drift (D)}$$

> *Note:* The crossing speed for a swimmer across a river may vary but is generally limited to 1 meter per second. All measurements must be in the same unit of measure (e.g., meters, feet).

The Buddy System

Whenever a Marine unit must enter into or operate on the water, a "buddy system" is employed in which every Marine is paired with a swimming partner. The buddy system matches an experienced swimmer with a weak swimmer. The experienced swimmer assists and encourages the weaker swimmer and bolsters confidence during night crossings. If a unit has an odd number of Marines, place the extra person with another pair to form a three-person team.

Water Crossings

Care of Weapons

Marine infantry weapons and munitions are designed to be able to operate after immersion. However, protect your weapons from moisture whenever possible.

A gas-operated weapon can malfunction if water travels down the barrel and enters the gas tube. To protect the gas tube—

Close the weapon's bolt before entering the water.

Seal the muzzle with a condom, balloon, plastic spoon wrapper, or other form of waterproof material.

Tie or melt the protective cover to create a watertight seal.

When the muzzle's protective cover is no longer needed, remove it. Open the bolt and inspect the barrel. If the tactical situation permits, swab excess moisture from the barrel. Test fire automatic weapons, if possible. Field strip and clean weapons as soon as possible. If time does not allow for a complete inspection, rinse inaccessible areas with small amounts of diesel fuel, then dry.

ISOMAT Raft

Construction of an ISOMAT raft is time-consuming. This type of raft should not be employed as part of an attack, but used for logistical purposes (e.g., evacuating stretcher cases, transporting supplies). Use the following steps to build an ISOMAT raft:

Wrap ISOMAT sleeping pads around sturdy sticks.

Use parachute cord and square knots to tie the pads securely in place and to lash stick ends together in a rectangle.

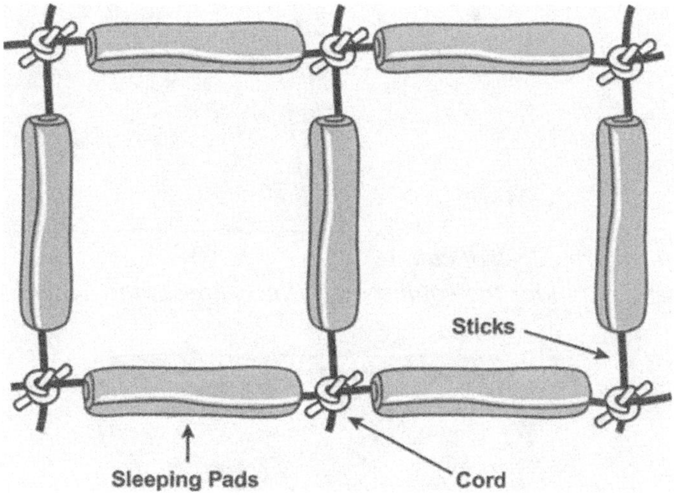

NOTE: The ISOMAT raft pictured can support several hundred pounds. However, the cargo will get wet if not properly waterproofed.

Poncho Raft

A poncho raft can support two Marines and their equipment and is well suited for long crossings. Use the following steps to build a poncho raft:

Inspect two ponchos and ensure they are serviceable.

Lay one poncho flat on the ground, with the hood-side up.

Cinch the hood tightly to form a gooseneck or tie in a knot.

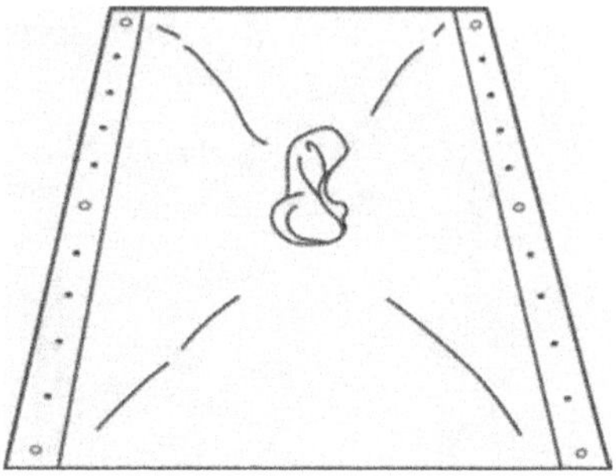

Pad sharp edges of equipment and place the equipment in the center of the poncho.

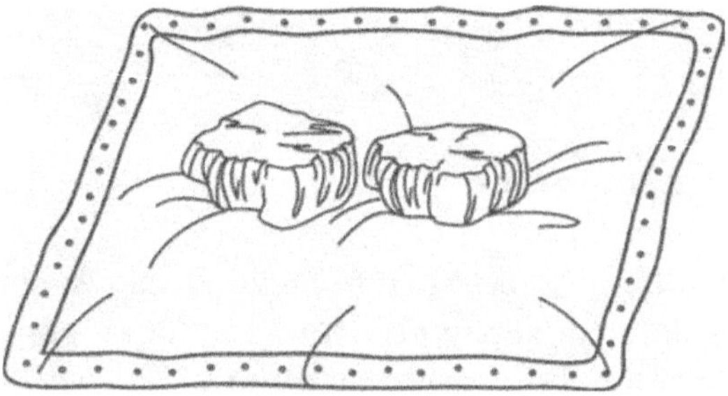

Place the second poncho over the equipment, rubber side up, and hood facing down.

Snap the edges of the two ponchos together.

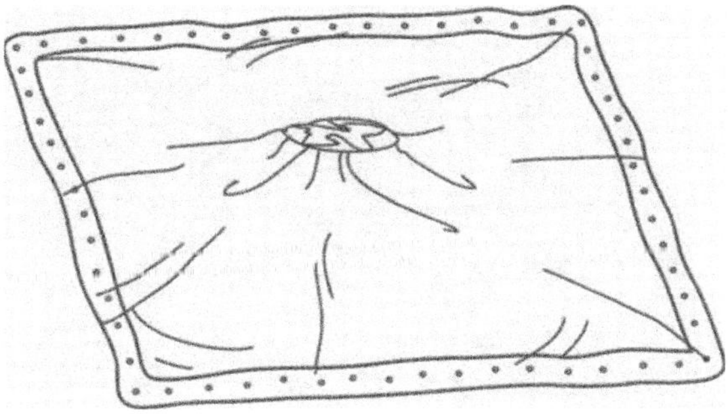

Roll the edges toward the equipment.

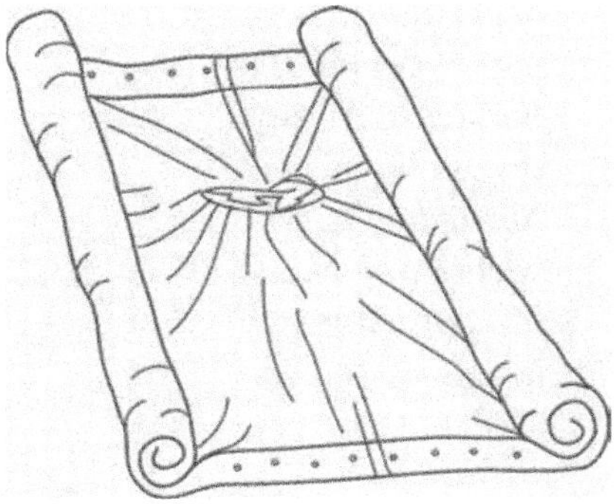

Roll the edges into pigtails and tie them off.

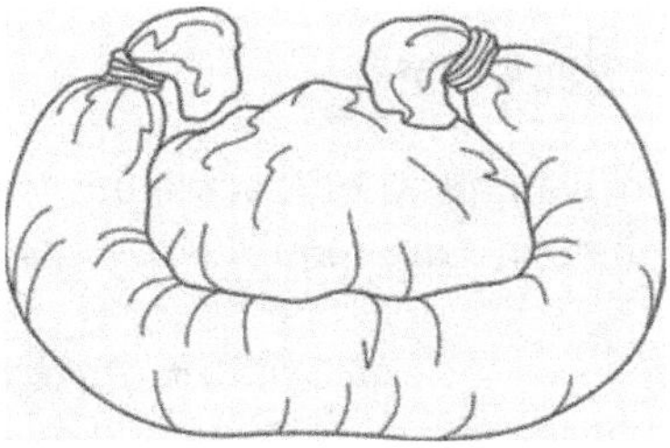

Pull the pigtails together over the top and lash them securely.

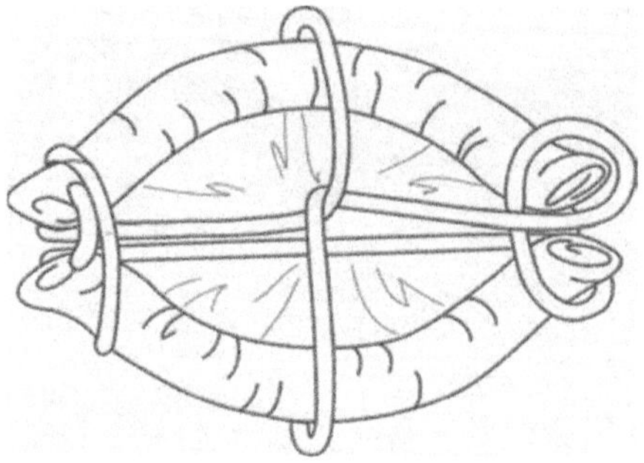

Protect the raft from brush punctures while placing it in the water. Swim across the water obstacle while security elements are covering the far shore.

Construction of a Pack Raft

You will need two waterproofed packs and two M16A2 service rifles. The following steps are required to construct a pack raft:

Place two packs side-by-side with the pack frames on the deck. The tops of the packs are opposite of each other.

Loosen the main compartment straps on both packs.

Insert one rifle on each end between the straps and the packs. The muzzles are opposite of each other. The rifles serve as one means to secure the packs together. Place the front sight post under the top flap.

Tighten the straps so that the rifles and packs are secure.

Take the excess strap on the inner side of each pack and secure it to the opposite pack to better secure the two packs together.

Take the excess straps on the outer sides of the packs and use those straps as safety lashing for the rifles.

Tuck the excess straps and check to make sure the rifles and packs are secure.

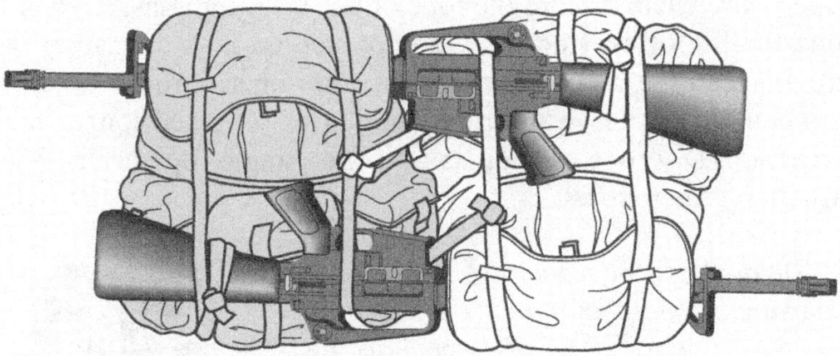

Single-Rope Bridge

A single-rope bridge offers a temporary and quick way to cross small rivers. It also provides extra security while crossing swift waters. At night, it prevents straggling, and guides units precisely from one side of the river to the other side. If crossing a river at night, plan for at least one single-rope bridge.

If your unit is crossing a river with swift currents or water depths above 4 feet, the unit is carrying sufficient rope to span the crossing site, and the tactical situation permits, secure the rope on near and far banks to provide a hand-hold for crossing Marines. This reduces the time required for the entire unit to cross and provides a degree of comfort/confidence for poor swimmers. Use a squad-sized bridge team to construct a single-rope bridge. Station several strong swimmers at the water's edge to help anyone who has trouble crossing.

Nylon rope is normally coiled in 120 foot lengths. It is 0.6 inches in diameter and has a breaking strength of about 3,840 pounds. Over time, a nylon rope can stretch to as much as one-third more than its original length and stretching weakens the rope. If the rope is stretched, discard the rope or use it for light tasks. To prolong the life of a nylon rope, do not step on it or drag it on the ground. Pad the rope in places where it contacts rocks or sharp corners. Do not leave the rope knotted or stretched longer than necessary. Dry rope as soon as possible. Single-rope bridge construction is as follows:

Tie a sling rope around your waist using a square knot and two, separate half hitches. See the appendix for detailed information on knots.

Attach a locking steel carabiner to the sling rope.

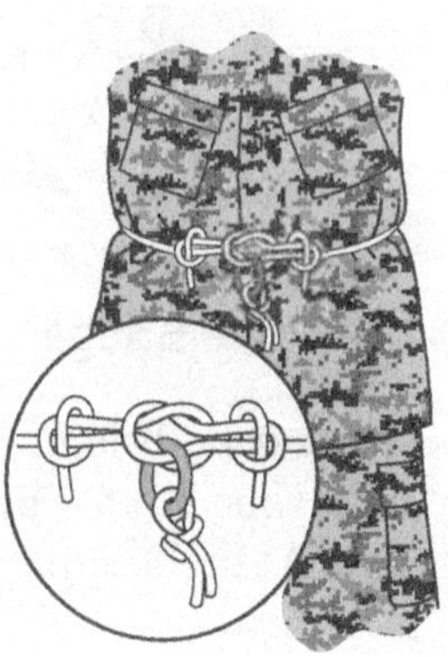

Tie a bowline knot in the running end of the bridge rope and attach it to the carabiner.

Temporarily secure the other end of the rope to a tree on the near shore.

Enter and cross the water.

> NOTE: Carry only your weapon and ammunition.

Exit the water on the opposite shore.

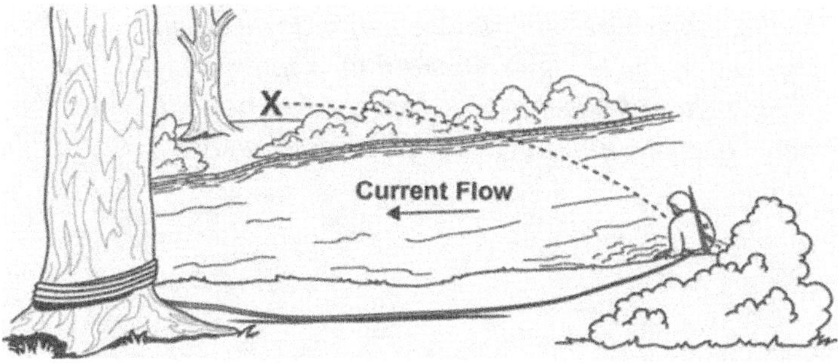

Prepare your weapon for use. Unhook the bridge rope from the carabiner at your waist, and tie the bridge rope to a sturdy tree using a round turn and two half hitches.

Conduct a box reconnaissance of the opposite shore.

On the near shore, have another Marine prepare to tighten the rope. That Marine should place a transport tightening system in the bridge rope by tying a double butterfly knot and placing two carabiners in the butterfly. See the appendix for detailed information on knots.

The Marine should pass the running end of the bridge rope around the downstream side of the near shore anchor point and through the two carabiners.

Pull the butterfly knot approximately one-third of the distance across the river.

Secure the bridge rope to an anchor point using a round turn and two half-hitches.

On the near shore, the Marine helping you should pull the slack out of the bridge rope until the butterfly knot is back on the near side. The bridge rope is then tied off against itself using two half hitches with a quick release in the last half hitch.

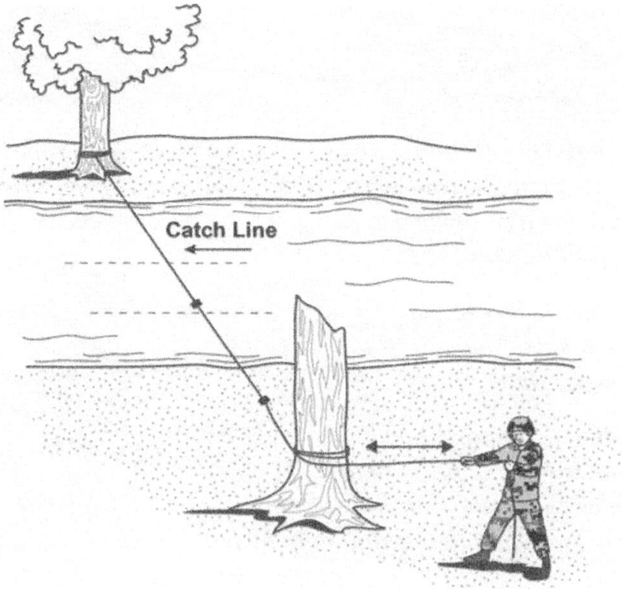

Note: The single-rope bridge must be as tight as possible so it will not sag when used.

If you lose your footing and fall into the water, swim with the current to the closest shore. Swimming against the current is dangerous and quickly causes fatigue.

High and Dry Crossings

If the single-rope bridge is high enough, suspend yourself below the single-rope bridge and above the water. Use the following steps to suspend yourself from a single-rope bridge and then pull yourself across the water:

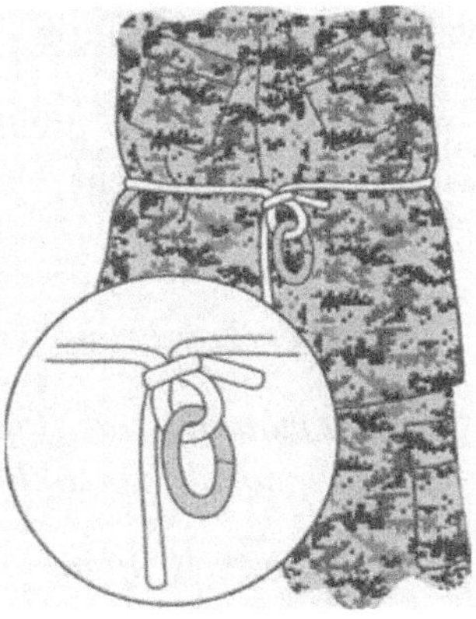

Tie a sling rope around your waist using a bowline. Ensure that the knot is tight.

Attach a carabiner through the bowline's loop. The carabiner's gate faces up.

Secure your helmet chin strap.

Face the single-rope bridge with your left shoulder toward the far shore.

Grasp the bridge rope in both hands.

Swing your body beneath the single-rope bridge with your head toward the far shore. Cross your ankles above the bridge rope.

Arch back until the carabiner contacts the bridge rope. Connect the carabiner to the bridge rope. Allow the carabiner to bear your body's weight.

Pull yourself across the single-rope bridge, hand over hand, to the far shore.

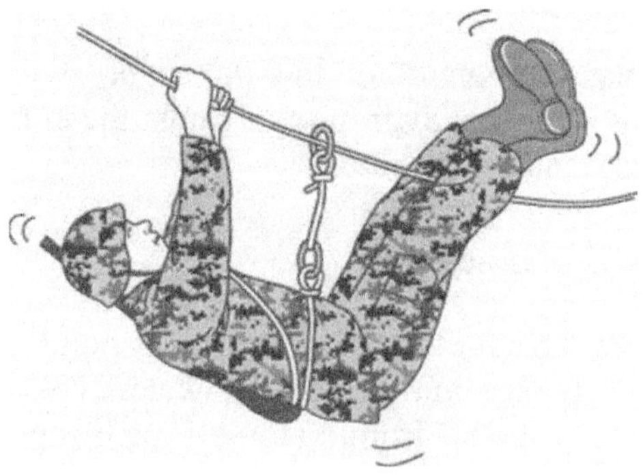

Swift Current Crossings

A single-rope bridge prevents being knocked down and swept away by a swift current. Use the following steps to move through a swift current:

Tie one end of a sling rope around your waist using a bowline.

Tie the running end of the sling rope in another bowline, and attach a carabiner to the bowline's loop.

Step up to the bridge. Face upstream.

Hook the carabiner to the single-rope bridge.

Walk sideways into the river while grasping the bridge rope in both hands.

Use the single-rope bridge for balance and remain standing, if possible.

Continue to move sideways through the river to the far shore.

Slow Current Crossings

If you face little or no current, it is not necessary to hook up to a bridge rope with a carabiner. Lie on your back in the water beneath the single-rope bridge. Support your body weight with your waterproof pack. Use the bridge rope and pull yourself hand over hand across the river.

Removal

If you are the last Marine waiting to cross, pull on the standing end of the rope to release the knot, then tie the rope around your waist using a bowline. The Marines on the far shore will pull you through the water.

Appendix
Knot Tying

Square Knot

The square knot is used to secure two ropes of equal diameter together so they form one continuous rope that will not slip.

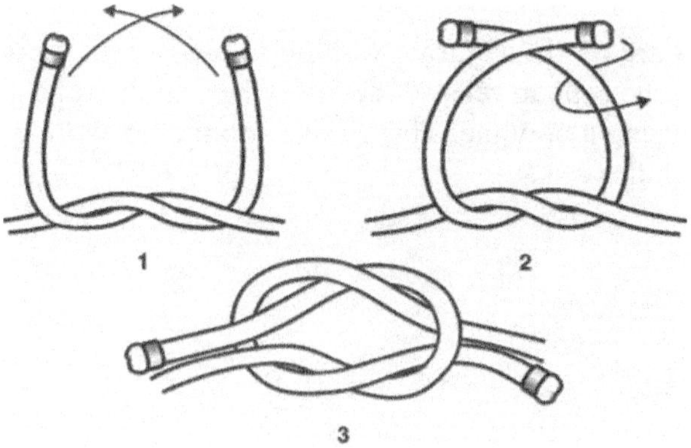

Bowline

The bowline forms a loop that will not tighten or slip under strain. It is easily untied.

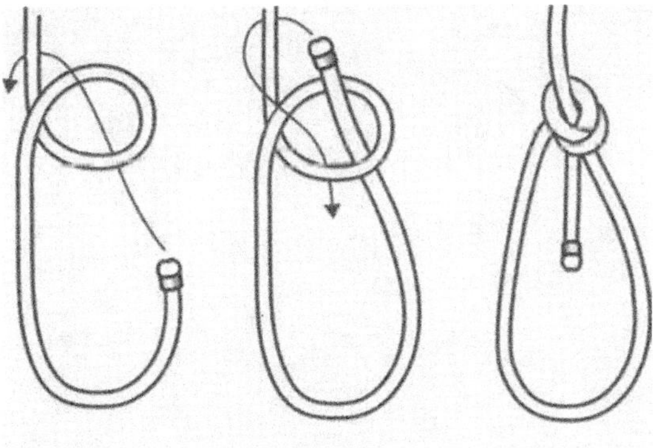

1 2 3

Hitches

Hitches are used to dress (prepare) knots and secure loose ends.

Half Hitch

The half hitch is used to tie a rope to a tree or to a larger rope. It will hold against a steady pull, but is not a secure hitch. It is frequently used to secure the free end of a rope.

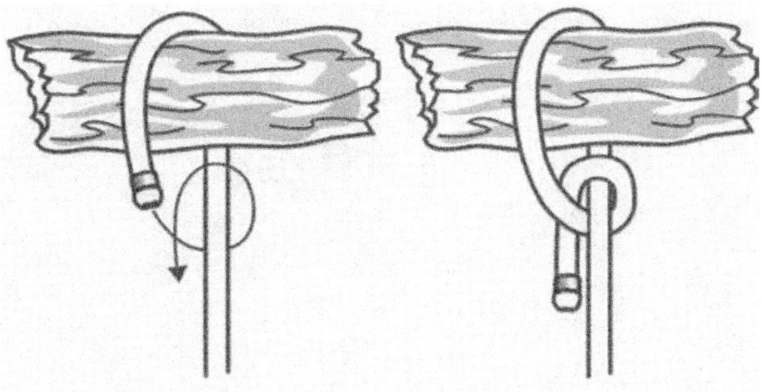

1 2

Two Half Hitches

Two half hitches can be used to secure the running end of a rope.

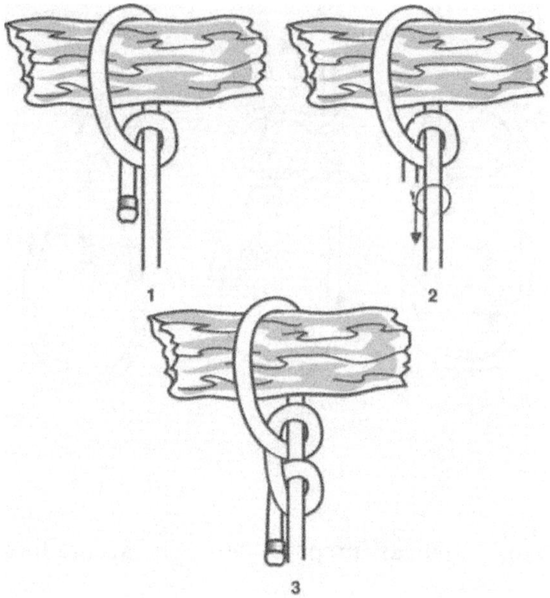

Round Turn and Two Half Hitches

A round turn and two half hitches can be used to fasten a rope to a tree. This hitch does not jam.

Butterfly Knot

The butterfly knot is used for anchor lines. Tied properly, it will not tighten on itself to the point that it cannot be easily untied. The butterfly knot can be used to pull a rope bridge taut. This knot can be used to tighten a fixed rope when mechanical means are not available. It will not jam if a stick is placed between the two upper loops.

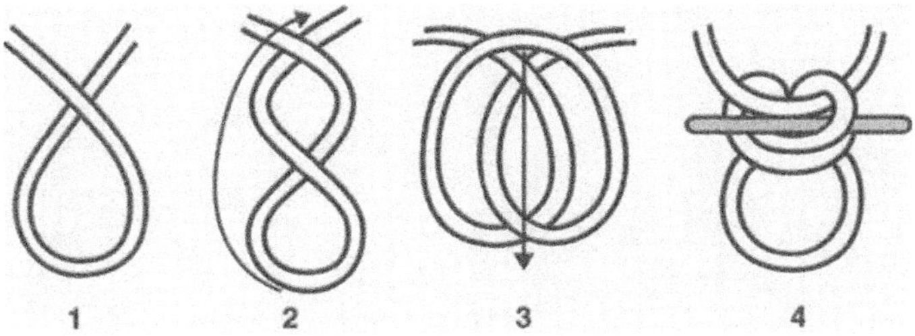

Jungle Operations

Department of Defense

Pentagon
Publishing

FIELD MANUAL
NO. 90-5

*FM 90-5

HEADQUARTERS
DEPARTMENT OF THE ARMY
Washington, DC, 16 August 1982

JUNGLE OPERATIONS

CONTENTS

*This manual supersedes FM 31-35, 26 September 1969.

The words "he," "him," "his," "man," and "men," when used in this publication, represent both the masculine and feminine genders unless otherwise specifically stated.

You may recommend changes to this manual to improve it. Key your comments to the page and line of text in which the changes are recommended. Give reasons for each comment to help understand and evaluate it. Send comments to: Commandant, United States Army Infantry School, ATTN: ATSH-B-ID, Fort Benning, Georgia 31905.

CHAPTER 1

The Jungle Environment

Section I. GENERAL

This chapter introduces jungle environments—where they are found and what they are like. Later chapters build on this information, providing guidance on fighting and living in the jungle.

Field Marshal Slim's words reflect the image of the jungle most armies carry into jungle warfare. At first, the jungle seems to be very hostile, but the hostility wanes as troops learn more about the jungle environment.

Jungles, in their various forms, are common in tropical areas of the world—mainly Southeast Asia, Africa, and Latin America.

CONTENTS

"To our men. . . the jungle was a strange, fearsome place; moving and fighting in it were a nightmare. We were too ready to classify jungle as 'impenetratable' . . . To us it appeared only as an obstacle to movement; to the Japanese it was a welcome means of concealed maneuver and suprise . . . The Japanese reaped the deserved reward . . . we paid the penalty."

-Field Marshall Slim, Victor in Burma, World War II (Concerning the dark, early days of the Burma Campaign)

JUNGLE REGIONS OF THE WORLD

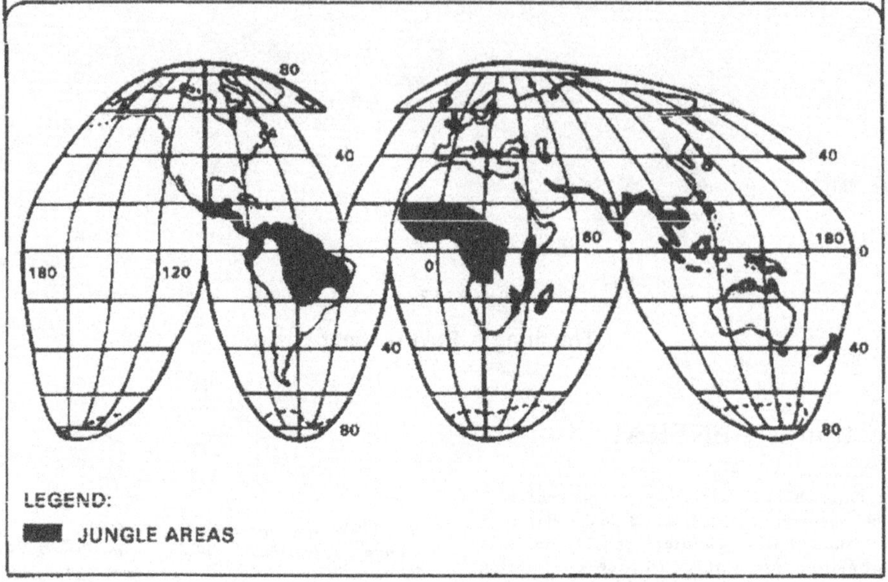

Section II. CLIMATE AND WEATHER.

The climate in jungles varies with location. Close to the equator, all seasons are nearly alike, with rains throughout the year; farther from the equator, especially in India and Southeast Asia, jungles have distinct wet (monsoon) and dry seasons. Both zones have high temperatures (averaging 78 to 95+ degrees Fahrenheit), heavy rainfall (as much as 1,000 centimeters [400+ inches] annually), and high humidity (90 percent) throughout the year.

Severe weather also has an impact on tactical operations in the jungle. The specific effects of weather on operations are discussed throughout this manual.

Jungle climates (high temperatures, high humidity, heavy rain) seriously affect:

- Men
- Clothing
- Equipment
- Weapons
- Vehicles
- Maintenance operations
- Training
- Tactics

Section III. TERRAIN AND VEGETATION

TYPES OF JUNGLES

The jungle environment includes densely forested areas, grasslands, cultivated areas, and swamps. Jungles are classified as primary or secondary jungles based on the terrain and vegetation.

PRIMARY JUNGLES

These are tropical forests. Depending on the type of trees growing in these forests, primary jungles are classified either as tropical rain forests or as deciduous forests.

Tropical Rain Forests. These consist mostly of large trees whose branches spread and lock together to form canopies. These canopies, which can exist at two or three different levels, may form as low as 10 meters from the ground. The canopies prevent sunlight from reaching the ground, causing a lack of undergrowth on the jungle floor. Extensive above-ground root systems and hanging vines are common. These conditions, combined with a wet and soggy surface, make vehicular traffic difficult. Foot movement is easier in tropical rain forests than in other types ofjungle. Except where felled trees or construction make a gap in the canopy of the rain forest, observation from the air is nearly impossible. Ground observation is generally limited to about 50 meters (55 yards).

TROPICAL RAIN FOREST

DECIDUOUS FOREST

Deciduous Forests. These are found in semitropical zones where there are both wet and dry seasons. In the wet season, trees are fully leaved; in the dry season, much of the foliage dies. Trees are generally less dense in deciduous forests than in rain forests. This allows more rain and sunlight to filter to the ground, producing thick undergrowth. In the wet season, with the trees in full leaf, observation both from the air and on the

ground is limited. Movement is more difficult than in the rain forest. In the dry season, however, both observation and trafficability improve.

SECONDARY JUNGLES

These are found at the edge of the rain forest and the deciduous forest, and in areas where jungles have been cleared and abandoned. Secondary jungles appear when the ground has been repeatedly exposed to sunlight. These areas are typically overgrown with weeds, grasses, thorns, ferns, canes, and shrubs. Foot movement is extremely slow and difficult. Vegetation may reach to a height of 2 meters. This will limit observation to the front to only a few meters.

COMMON JUNGLE FEATURES

SWAMPS

These are common to all low jungle areas where there is water and poor drainage. There are two basic types of swamps—mangrove and palm.

Mangrove Swamps. These are found in coastal areas wherever tides influence water flow. The mangrove is a shrub-like tree which grows 1 to 5 meters high. These trees have tangled root systems, both above and below the water level, which restrict movement to foot or small boats. Observation in mangrove swamps, both on the ground and from the air, is poor. Concealment is excellent.

Palm Swamps. These exist in both salt and fresh water areas. Like movement in the mangrove swamps, movement through palm swamps is mostly restricted to foot (sometimes small boats). Vehicular traffic is nearly impossible except after extensive road construction by engineers. Observation and fields-of-fire are very limited. Concealment from both air and ground observation is excellent.

SAVANNA

This is a broad, open jungle grassland in which trees are scarce. The thick grass is broad-bladed and grows 1 to 5 meters high. Movement in the savanna is generally easier than in other types of jungle areas, especially for vehicles. The sharp-edged, dense grass and extreme heat make foot movement a slow and tiring process. Depending on the height of the grass, ground observation may vary from poor to good. Concealment from air observation is poor for both troops and vehicles.

BAMBOO

This grows in clumps of varying size in jungles throughout the tropics. Large stands of bamboo are excellent obstacles for wheeled or tracked vehicles. Troop movement through bamboo is slow, exhausting, and noisy. Troops should bypass bamboo stands if possible.

CULTIVATED AREAS

These exist in jungles throughout the tropics and range from large, well-planned and well-managed farms and plantations to small tracts cultivated by individual farmers. There are three general types of cultivated areas—rice paddies, plantations, and small farms.

Rice Paddies. These are flat, flooded fields in which rice is grown. Flooding of the fields is controlled by a network of dikes and irrigation ditches which make movement by vehicles difficult even when the fields are dry. Concealment is poor in rice paddies. Cover is limited to the dikes, and then only from ground fire. Observation and fields of fire are excellent. Foot movement is poor when the fields are wet because soldiers must

wade through water about 1/2 meter (2 feet) deep and soft mud. When the fields are dry, foot movement becomes easier. The dikes, about 2 to 3 meters tall, are the only obstacles.

Plantations. These are large farms or estates where tree crops, such as rubber and coconut, are grown. They are usually carefully planned and free of undergrowth (like a well-tended park). Movement through plantations is generally easy. Observation along the rows of trees is generally good. Concealment and cover can be found behind the trees, but soldiers moving down the cultivated rows are exposed.

Small Farms. These exist throughout the tropics. These small cultivated areas are usually hastily planned. After 1 or 2 years' use, they usually are abandoned, leaving behind a small open area which turns into secondary jungle. Movement through these areas may be difficult due to fallen trees and scrub brush.

Generally, observation and fields-of-fire are less restricted in cultivated areas than in uncultivated jungles. However, much of the natural cover and concealment are removed by cultivation, and troops will be more exposed in these areas.

CHAPTER 2

Life in the Jungle

Section I. GENERAL

Soldiers must understand that the environment affects everyone. The degree to which soldiers are trained to live and fight in harsh environments will determine their unit's success or failure.

"Jungle fighting is not new to US soldiers, nor does the enemy have a monopoly on jungle know-how. US units adapted well to jungle fighting, and when we operated against the North Vietnamese Army along the Cambodian border we found that they had as much difficulty operating in the area as we did. The prisoners we captured were, as a rule, undernourished, emaciated, and sick with malaria. They stated that almost everyone in their unit had malaria, and many had died from it."

Report, 25th Infantry Division,
Republic of Vietnam

There is very little to fear from the jungle environment. Fear itself can be an enemy. Soldiers must be taught to control their fear of the jungle. A man overcome with fear is of little value in any situation. Soldiers in a jungle must learn that the most important thing is to keep their heads and calmly think out any situation.

Many of the stories written about out-of-the-way jungle places were written by writers who went there in search of adventure rather than facts. Practically without exception, these authors exaggerated or invented many of the thrilling experiences they relate. These thrillers are often a product of the author's imagination and are not facts.

Most Americans, especially those raised in cities, are far removed from their pioneer ancestors, and have lost the knack of taking care of themselves under all conditions. It would be foolish to say that, without proper training, they would be in no danger if lost in the jungles of Southeast Asia, South America, or some Pacific island. On the other hand, they would be in just as much danger if lost in the mountains of western Pennsylvania or in other undeveloped regions of our own country. The only difference would be that a man is less likely to panic when he is lost in his homeland than when he is lost abroad.

CONTENTS

Section II. JUNGLE HAZARDS

EFFECT OF CLIMATE

The discomforts of tropical climates are often exaggerated, but it is true that the heat is more persistent. In regions where the air contains a lot of moisture, the effect of the heat may seem worse than the same temperature in a dry climate. Many people experienced in jungle operations feel that the heat and discomfort in some US cities in the summertime are worse than the climate in the jungle.

Strange as it may seem, there may be more suffering from cold in the tropics than from the heat. Of course, very low temperatures do not occur, but chilly days and nights are common. In some jungles, in winter months, the nights are cold enough to require a wool blanket or poncho liner for sleeping.

Rainfall in many parts of the tropics is much greater than that in most areas of the temperate zones. Tropical downpours usually are followed by clear skies, and in most places the rains are predictable at certain times of the day. Except in those areas where rainfall may be continuous during the rainy season, there are not many days when the sun does not shine part of the time.

People who live in the tropics usually plan their activities so that they are able to stay under shelter during the rainy and hotter portions of the day. After becoming used to it, most tropical dwellers prefer the constant climate of the torrid zones to the frequent weather changes in colder climates.

INSECTS

Malaria-carrying mosquitoes are probably the most harmful of the tropical insects. Soldiers can contract malaria if proper precautions are not taken.

Precautions against malaria include:

- Taking Dapsone and chloroquine-primaquine
- Using insect repellent
- Wearing clothing that covers as much of the body as possible
- Using nets or screens at every opportunity
- Avoiding the worst-infested areas when possible

PROTECTION AGAINST MALARIA

Mosquitoes are most prevalent early at night and just before dawn. Soldiers must be especially cautious at these times. Malaria is more common in populated areas than in uninhabited jungle, so soldiers must also be especially cautious when operating around villages. Mud packs applied to mosquito bites offer some relief from itching.

Wasps and bees may be common in some places, but they will rarely attack unless their nests are disturbed. When a nest is disturbed, the troops must leave the area and reassemble at the last rally point. In case of stings, mud packs are helpful. In some areas, there are tiny bees, called sweatbees, which may collect on exposed parts of the body during dry weather, especially if the body is sweating freely. They are annoying but stingless and will leave when sweating has completely stopped, or they may be scraped off with the hand.

The larger centipedes and scorpions can inflict stings which are painful but not fatal. They like dark places, so it is always advisable to shake out blankets before sleeping at night, and to make sure before dressing that they are not hidden in clothing or shoes. Spiders are commonly found in the jungle. Their bites may be painful, but are rarely serious. Ants can be dangerous to injured men lying on the ground and unable to move. Wounded soldiers should be placed in an area free of ants.

In Southeast Asian jungles, the rice-borer moth of the lowlands collects around lights in great numbers during certain seasons. It is a small, plain-colored moth with a pair of tiny black spots on the wings. It should never be brushed off roughly, as the small barbed hairs of its body may be ground into the skin. This causes a sore, much like a burn, that often takes weeks to heal.

LEECHES

Leeches are common in many jungle areas, particularly throughout most of the Southwest Pacific, Southeast Asia, and the Malay Peninsula. They are found in swampy areas, streams, and moist jungle country. They are not poisonous, but their bites may become infected if not cared for properly. The small wound that they cause may provide a point of entry for the germs which cause tropical ulcers or "jungle sores." Soldiers operating in the jungle should watch for leeches on the body and brush them off before they have had time to bite. When they have taken hold, they should not be pulled off forcibly because part of the leech may remain in the skin. Leeches will release themselves if touched with insect repellent, a moist piece of tobacco, the burning end of a cigarette, a coal from a fire, or a few drops of alcohol.

Straps wrapped around the lower part of the legs ("leech straps") will prevent leeches from crawling up the legs and into the crotch area. Trousers should be securely tucked into the boots.

LEECH STRAPS

SNAKES

A soldier in the jungle probably will see very few snakes. When he does see one, the snake most likely will be making every effort to escape.

If a soldier should accidently step on a snake or otherwise disturb a snake, it will probably attempt to bite. The chances of this happening to soldiers traveling along trails or waterways are remote if soldiers are alert and careful. Most jungle areas pose less of a snakebite danger than do the uninhabited areas of New Mexico, Florida, or Texas. This does not mean that soldiers should be careless about the possibility of snakebites, but ordinary precautions against them are enough. Soldiers should be particularly watchful when clearing ground.

Treat all snakebites as poisonous.

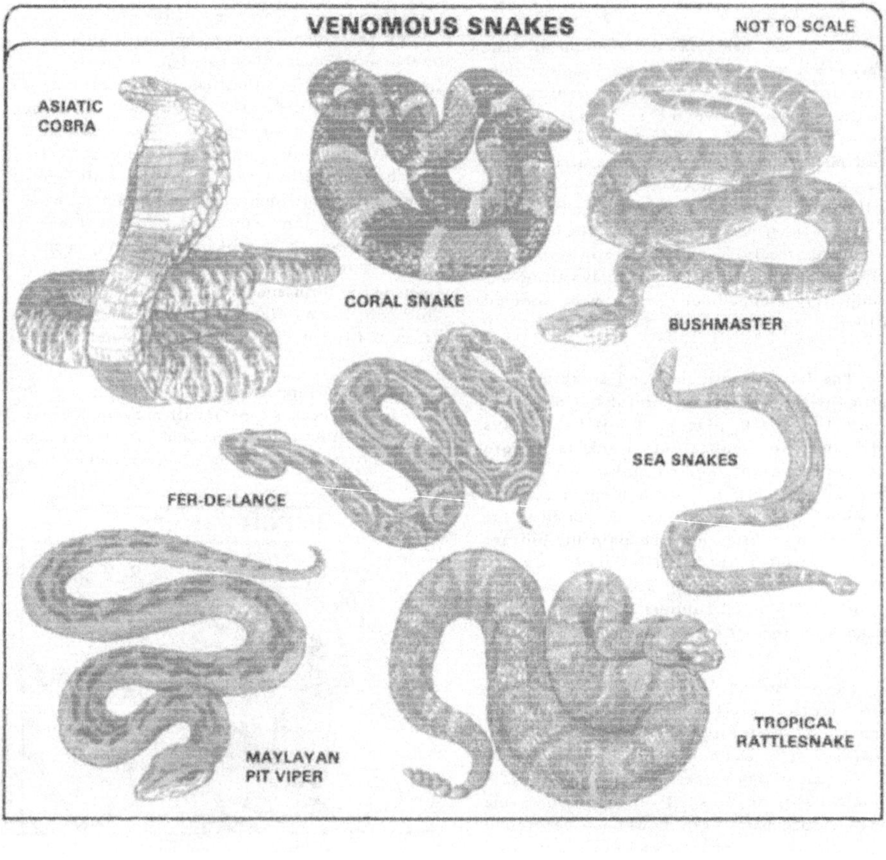

SNAKEBITE TREATMENT

Follow these steps if bitten:

- Remain calm, but act swiftly, and chances of survival are good. (Less than one percent of properly treated snakebites are fatal. Without treatment, the fatality rate is 10 to 15 percent.)
- Immobilize the affected part in a position below the level of the heart.
- Place a lightly constricting band 5 to 10 centimeters (2 to 4 inches) closer to the heart than the site of the bite. Reapply the constricting band ahead of the swelling if it moves up the arm or leg. The constricting band should be placed tightly enough to halt the flow of blood in surface vessels, but not so tight as to stop the pulse.
- Do not attempt to cut open the bite or suck out venom.
- Seek medical help. If possible, the snake's head with 5 to 10 centimeters (2 to 4 inches) of its body attached should be taken to the medics for identification. Identification insures use of the proper antivenom.

CROCODILES AND CAYMANS

Crocodiles and caymans are meat-eating reptiles which live in tropical areas. "Crocodile-infested rivers and swamps" is a catch-phrase often found in stories about the tropics. Asian jungles certainly have their share of crocodiles, but there are few authenticated cases of crocodiles actually attacking humans. Caymans, found in South and Central America, are not likely to attack unless provoked.

CROCODILES AND CAYMANS

WILD ANIMAL

In Africa, where lions, leopards, and other flesh-eating animals abound, they are protected from hunters by local laws and live on large preserves. In areas where the beasts are not protected, they are shy and seldom seen. When encountered, they will attempt to escape. All large animals can be dangerous if cornered or suddenly startled at close quarters. This is especially true of females with young. In the jungles of Sumatra, Bali, Borneo, Southeast Asia, and Burma there are tigers, leopards, elephants, and buffalo. Latin America's jungles have the jaguar. Ordinarily, these will not attack a man unless they are cornered or wounded.

Certain jungle animals, such as water buffalo and elephants, have been domesticated by the local people. Soldiers should also avoid these animals. They may appear tame, but this tameness extends only to people the animals are familar with.

POISONOUS VEGETATION

Another area of danger is that of poisonous plants and trees. For example, nettles, particularly tree nettles, are one of the dangerous items of vegetation. These nettles have a severe stinging that will quickly educate the victim to recognize the plant. There are ringas trees in Malaysia which affect some people in much the same way as poison oak. The poison ivy and poison sumac of the continental US can cause many of the same type troubles that may be experienced in the jungle. The danger from poisonous plants in the woods of the US eastern seaboard is similar to that of the tropics. Thorny thickets, such as rattan, should be avoided as one would avoid a blackberry patch.

Some of the dangers associated with poisonous vegetation can be avoided by keeping sleeves down and wearing gloves when practical.

HEALTH AND HYGIENE

The climate in tropical areas and the absence of sanitation facilities increase the chance that soldiers may contract a disease. Disease is fought with good sanitation practices and preventive medicine. In past wars, diseases accounted for a significantly high percentage of casualties.

Before going into a jungle area, leaders must:

- **Make sure immunizations are current.**
- **Get soldiers in top physical shape.**
- **Instruct soldiers in personal hygiene.**

Upon arrival in the jungle area, leaders must:

- **Allow time to adjust (acclimate) to the new environment.**
- **Never limit the amount of water soldiers drink. (It is very important to replace the fluids lost through sweating.)**
- **Instruct soldiers on the sources of disease. Insects cause malaria, yellow fever, and scrub typhus. Typhoid, dysentery, cholera, and hepatitis are caused by dirty food and contaminated water.**

WATERBORNE DISEASES

Water is vital in the jungle and is usually easy to find. However, water from natural sources should be considered contaminated. Water purification procedures must be taught to all soldiers. Germs of serious diseases, like dysentery, are found in impure water. Other waterborne diseases, such as blood fluke, are caused by exposure of an open sore to impure water.

Soldiers can prevent waterborne diseases by:

- **Obtaining drinking water from approved engineer water points.**
- **Using rainwater; however, rainwater should be collected after it has been raining at least 15 to 30 minutes. This lessens the chances of impurity being washed from the jungle canopy into the water container. Even then the water should be purified.**
- **Insuring that all drinking water is purified.**
- **Not swimming or bathing in untreated water.**
- **Keeping the body fully clothed when crossing water obstacles.**

FUNGUS DISEASES

These diseases are caused by poor personal health practices. The jungle environment promotes fungus and bacterial diseases of the skin and warm water immersion skin diseases. Bacteria and fungi are tiny plants which multiply fast under the hot, moist conditions of the jungle. Sweat-soaked skin invites fungus attack. The following are common skin diseases that are caused by long periods of wetness of the skin:

Warm Water Immersion Foot. This disease occurs usually where there are many creeks, streams, and canals to cross, with dry ground in between. The bottoms of the feet become white, wrinkled, and tender. Walking becomes painful.

Chafing. This disease occurs when soldiers must often wade through water up to their waists, and the trousers stay wet for hours. The crotch area becomes red and painful to even the lightest touch.

Most skin diseases are treated by letting the skin dry.

To prevent these diseases, soldiers should:

- **Bathe often, and air- or sun-dry the body as often as possible.**
- **Wear clean, dry, loose-fitting clothing whenever possible.**
- **Not sleep in wet, dirty clothing. Soldiers should carry one dry set of clothes just for sleeping. Dirty clothing, even if wet, is put on again in the morning. This practice not only fights fungus, bacterial, and warm water immersion diseases but also prevents chills and allows soldiers to rest better.**
- **Not wear underwear during wet weather. Underwear dries slower than jungle fatigues, and causes severe chafing.**
- **Take off boots and massage feet as often as possible.**
- **Dust feet, socks, and boots with foot powder at every chance.**
- **Always carry several pairs of socks and change them frequently.**
- **Keep hair cut short.**

HEAT INJURIES

These result from high temperatures, high humidity, lack of air circulation, and physical exertion. All soldiers must be trained to prevent heat disorders.

HEAT INJURIES			
TYPE	CAUSE	SYMPTOMS	TREATMENT
Dehydration	Dehydration is caused by the loss of too much water. About two-thirds of the human body is water. When water is not replaced as it is lost, the body becomes dried out—dehydrated.	The symptoms are sluggishness and listlessness.	The treatment is to give the victim plenty of water.
Heat Exhaustion	Heat exhaustion is caused by the loss of too much water and salt.	The symptoms are: Dizziness. Nausea. Headache. Cramps. Rapid, weak pulse. Cool, wet skin.	The treatment consists of: Moving the victim to a cool, shaded place for rest. Loosening the clothing. Elevating the feet to improve circulation. Giving the victim cool salt water (two salt tablets dissolved in a canteen of water). Natural sea water should not be used.
Heat Cramps	Heat cramps are caused by the loss of too much salt.	The symptom is painful muscle cramps which are relieved as soon as salt is replaced.	The treatment is the same as for heat exhaustion.
Heatstrokes	Heatstroke (sunstroke) is caused by a breakdown in the body's heat control mechanism. The most likely victims are those who are not acclimated to the jungle, or those who have recently had bad cases of diarrhea. Heatstroke can kill if not treated quickly.	The symptoms are: Hot, red, dry skin (most important sign). No sweating (when sweating would be expected). Very high temperature (105 to 110 degrees). Rapid pulse. Spots before eyes. Headache, nausea, dizziness, mental confusion. Sudden collapse.	Treatment consists of: Cooling the victim immediately. This is achieved by putting him in a creek or stream; pouring canteens of water over him; fanning him; and using ice, if available. Giving him cool salt water (prepared as stated earlier) if he is conscious. Rubbing his arms and legs rapidly. Evacuating him to medical aid as soon as possible.

> Heat injuries are prevented by:
>
> - Drinking plenty of water.
> - Using extra salt with food and water.
> - Slowing down movement.

NOTE: For more details, see FM 21-10 for field hygiene and sanitation, and FM 21-11 for first aid for soldiers.

NATIVES

Like all other regions of the world, the jungle also has its native inhabitants. Soldiers should be aware that some of these native tribes can be hostile if not treated properly.

There may be occasions, however, when hostile tribes attack without provacation. If they attack, a small force should be able to disperse them.

> To prevent a conflict, leaders should insure that their soldiers:
>
> - Respect the natives' privacy and personal property
> - Observe the local customs and taboos
> - Do not enter a native house without being invited
> - Do not pick fruits or cut trees without permission of their owners
> - Treat the natives as friends

Section III. JUNGLE SURVIVAL

FOOD

Food of some type is always available in the jungle—in fact, there is hardly a place in the world where food cannot be secured from plants and animals. All animals, birds, reptiles, and many kinds of insects of the jungle are edible. Some animals, such as toads and salamanders, have glands on the skin which should be removed before their meat is eaten. Fruits, flowers, buds, leaves, bark, and often tubers (fleshy plant roots) may be eaten. Fruits eaten by birds and monkeys usually may be eaten by man.

Meats that can be found in most jungles include:	The following types of fruits and nuts are common in jungle areas:		Vegetables found in most jungles include:
Wild fowl	Bananas	Wild raspberries	Taro *
Wild cattle	Coconuts	Nakarika	Yam *
Wild pig	Oranges and lemons	Papaya	Yucca *
Freshwater fish *	Navele nuts	Mangoes	Hearts of palm trees
Saltwater fish	Breadfruit		
Fresh-water crawfish			

*These items must be cooked before eating.

There are various means of preparing and preserving food found in the jungle. Fish, for example, can be cleaned and wrapped in wild banana leaves. This bundle is then tied with string made from bark, placed on a hastily constructed wood griddle, and roasted thoroughly until done. Another method is to roast the bundle of fish underneath a pile of red-hot stones.

Other meats can be roasted in a hollow section of bamboo, about 60 centimeters (2 feet) long. Meat cooked in this manner will not spoil for three or four days if left inside the bamboo stick and sealed.

Yams, taros, yuccas, and wild bananas can be cooked in coals. They taste somewhat like potatoes. Palm hearts can make a refreshing salad, and papaya a delicious dessert.

SHELTER

Jungle shelters are used to protect personnel and equipment from the harsh elements of the jungle. Shelters are necessary while sleeping, planning operations, and protecting sensitive equipment.

> **When selecting shelter, leaders should:**
>
> - **Choose high ground, away from swamps and dry river beds**
> - **Avoid trails, game tracks, or villages**
>
> (See app E for more on shelters.)

NAVIGATION

Navigation in thick jungle areas is difficult even for the most experienced navigators. Soldiers navigating in the jungle must use various aids. The compass is an obvious aid, but a soldier would never be able to move very fast in the jungle if he had to constantly move along a magnetic azimuth. Movement along a terrain feature, such as a ridgeline, is easier but can be extremely dangerous when establishing a pattern of consistency. A soldier must trust the compass, map, and pace count. A soldier should not keep his eyes riveted on the compass; however, it should be used as a check.

The shadows caused by the sun are an easily observed and accurate aid to direction. Allowances must be made for the gradual displacement of the shadows as the sun moves across the sky.

Other aids to maintaining direction include prominent objects, the course of rivers, prevailing winds, the stars, and the moon. *(See app B for more on navigation.)*

OBSERVATION

All movements of animals and men are marked by tracks and signs. Soldiers must learn to read signs left in soft ground, in streambeds, on roads and trails, and near watering places and salt licks. Animals seldom move without a reason; a few fresh tracks supply information about their maker, his direction, and probable intentions.

Animals avoid man. The animals, their tracks, and their behavior can reveal whether or not men are in the area. Jungle fighters can listen to the cries of animals and learn to recognize their alarm calls.

The ability to track and to recognize signs in the jungle are valuable skills. Throughout the soldier's time in the jungle, he should practice these skills. *(See app B for more on tracking.)*

CLOTHING AND EQUIPMENT

Before deploying for jungle operations, troops are issued special uniforms and equipment. Some of these items are:

JUNGLE FATIGUES

These fatigues are lighter and faster drying than standard fatigues. To provide the best ventilation, the uniform should fit loosely. It should never be starched.

JUNGLE BOOTS

These boots are lighter and faster drying than all-leather boots. Their cleated soles will maintain footing on steep, slippery slopes. The ventilating insoles should be washed in warm, soapy water when the situation allows.

INSECT (MOSQUITO) BAR

The insect (mosquito) bar or net should be used any time soldiers sleep in the jungle. Even if conditions do not allow a shelter, the bar can be hung inside the fighting position or from trees or brush. No part of the body should touch the insect net when it is hung, because mosquitoes can bite through the netting. The bar should be tucked or laid loosely, not staked down. Although this piece of equipment is very light, it can be bulky if not folded properly. It should be folded inside the poncho as tightly as possible.

CLOTHING AND EQUIPMENT

CHAPTER 3

Preparation and Training to Deploy to Jungle Areas

Section I. GENERAL

Chapters 1 and 2 describe the jungle environment. Since many soldiers are unaccustomed to such an environment, they must make preparations before conducting jungle operations.

This chapter lists the factors to be considered when preparing for jungle operations and presents training tips for conducting jungle training.

Section II. FACTORS TO BE CONSIDERED WHEN PREPARING FOR JUNGLE OPERATIONS

When a unit is alerted for training or actual combat operations in a jungle environment, the commander must first consider the following:

- Where will the unit be training or operating?
- What are the climatic and terrain conditions of the objective area?
- How much time does the unit have to prepare?
- What available training areas have climate and terrain resembling the objective area?
- What type operations are to be conducted—conventional or counterguerrilla?

CONTENTS

- Will the unit be taking its own equipment?
- Does any of the unit's equipment require modification (including camouflage painting)?
- What special equipment does the unit require?
- Does the unit have any jungle warfare instructors, soldiers with jungle experience, or linguists?
- What training assistance is available? Outside instructors? Training aids?
- Does higher headquarters have special standing operating procedures (SOP) for jungle war?
- Are all soldiers physically fit?
- What information is available about the enemy?
- What information is available about local civilians and allied forces in the objective area?

Once these questions have been answered, the commander can develop a program to prepare his unit to operate in the jungle.

The idea that a unit is technically and tactically proficient is only a small part of a unit's preparation. Emphasis should be placed on the mental, physical, and psychological aspects of operating in a jungle environment. The key to overcoming these problems lies with the unit chain of command in their efforts to develop a "will to win" as well as "will to train to win."

Section III. TRAINING TIPS

Units committed to jungle operations may have to fight as soon as they arrive in the operational area. Commanders must make the best use of the preparation time available. Measures which commanders should consider include:

Making use of time in garrison. Certain jungle subjects can be taught using classroom instruction. This training should begin as soon as possible, so that time in jungle training areas can be devoted to more advanced techniques.

Making use of local training areas. Although these training areas may not resemble jungle terrain, some jungle techniques can be introduced in them. This will provide a training base which can be expanded when the unit deploys to its jungle training or operational areas. In addition, physical training should begin in the local training area as early as possible before deployment.

Integrating individual training into unit training exercises. Rather than devote field training time to the individual skills required to live in the jungle, these skills should be introduced early in classes, and then practiced during unit training exercises.

The following lists can be used as a guide to subjects that should be covered:

INDIVIDUAL TRAINING

Common Subjects:

- Jungle environment and acclimation
- Living in the jungle
- Survival, evasion, and escape
- Camouflage and concealment
- Tracking
- Operational area orientation
- Jungle navigation
- Equipment recognition
- Enemy orientation
- Physical conditioning

Staff and Leader Subjects:

- Jungle maintenance and supply techniques
- Jungle terrain appreciation
- Enemy order of battle and tactics
- Airmobile techniques

Specialist Subjects:

- Language
- Demolitions
- Field expedient antennas
- Medevac techniques

UNIT TRAINING

- Movement
- Obstacles and Barriers
- Scouting, Surveillance, Patrolling, and Tracking
- Air Defense
- Adjustment and Conduct of Fires
- Immediate Action Drill
- Communications
- Jungle Operations:
  - Ambush/counterambush
  - Raid
  - Attack
  - Defense
  - Infiltration
  - Airmobile operations
  - Waterborne operations
  - River crossing
  - Road clearing

FIRE TEAM AND CREWMEMBERS

- Identification and Marking of Mines and Booby-traps
- Working with Helicopters
- Weapons Training

ACCLIMATION

The first priority in preparation for jungle warfare is acclimation (getting accustomed to jungle climate). Troops who are not conditioned properly will not perform jungle warfare tasks reliably. Different people become acclimated to hot weather at different rates, but the following methods can be used in most units.

Exercise is the best method for acclimation, because troops in good physical condition will adapt easily to new climates. A 7- to 14-day conditioning period should be sufficient for most soldiers. Exercises should be moderately strenuous at the beginning, and become more demanding each day. Troops from warm climates will adapt faster than troops deploying from colder climates. Physical training in heated gymnasiums prior to deployment will also ease the acclimation process.

Leaders must be alert for symptoms of heat disorders during the acclimation period.

SURVIVAL, EVASION, AND ESCAPE

Convincing a soldier that he will survive alone in the jungle will go a long way in building his self-confidence. FM 21-76 contains details on survival, evasion, and escape training.

SWIMMING

Swimming is also a vital skill for the jungle fighter. Falling into a jungle pool or river can be a terrible experience, especially for a nonswimmer. All troops should be "drown proofed" as shown in FM 21-20. Units should identify their strong swimmers for lifeguard training and other more difficult swimming tasks.

CAMOUFLAGE AND CONCEALMENT

Training to conceal soldiers and equipment from ground and air observation is equally important to combat, combat support, and combat service support units. Proper use of camouflage will help to make up for an enemy's superior knowledge of the jungle area. Appendix E contains information about jungle camouflage and concealment techniques.

JUNGLE LIVING

Following a short period of classroom instruction, soldiers should experience jungle living conditions in the field. This training can be incorporated into other unit training.

Subjects which should be stressed include:

- Heat disorders
- Survival
- First aid
- Health, hygiene, and field sanitation
- Proper wearing of clothing
- Use of equipment in a jungle environment
- Prevention and treatment of snakebites and insect bites

During this period of training, use of garrison facilities should be kept to a minimum. Supplies should be brought to the field rather than the unit returning to the rear for them. Soldiers should learn to live without unnecessary personal comforts.

Land navigation should be practiced using jungle movement techniques. *(See app B.)*

TARGET AREA ORIENTATION

Classes on the host country should stress those facts which apply to operations.

Subjects could include:

- Terrain appreciation
- Climate
- Population and culture
- Language (phrase books may be issued)
- Road, railroad, and canal system
- Standards of conduct for US Army personnel
- Allied armed forces
- Reasons for US involvement

JUNGLE TACTICS

Chapter 5 describes tactics common to jungle fighting. These tactics should be taught first to leaders down to squad level. The leaders then train their own units. Stress should be placed on small unit tactics and operations with Army aviation. Since night operations, especially ambushes, are common in jungle fighting, units should emphasize night training.

MOVEMENT

Units should train in tactical marches.

Training should emphasize:

- Breaching of obstacles
- Scouting, patrolling, and tracking
- Off-road movement over rugged terrain
- Air defense and counterambush drills

WEAPONS TRAINING

Most jungle fighting takes place at close range. Soldiers should be trained in "quick fire," as outlined in *chapter 8, FM 23-9* Advanced training should be conducted on a "jungle range." On this range, soldiers move down a trail and engage pop-up and moving targets which appear suddenly at close range. Targets are operated by an assistant on signals from a lane grader. Targets should be exposed for 3 to 6 seconds. Only 2 or 3 rounds should be fired at each target. At a later stage, boobytraps and obstacles can be emplaced on the trail.

INTELLIGENCE

Soldiers should be trained in specific intelligence subjects.

These include:

- Enemy organization and tactics
- Equipment recognition
- National markings
- Sound or signature recognition

This last subject is particularly important in jungle operations, because soldiers will more often hear weapons firing than see them. If captured enemy weapons and equipment are available, they should be used as aids in this training.

MAINTENANCE AND LOGISTICS

Chapter 6 describes some of the jungle's effects on equipment. Operators need to learn techniques to keep their equipment operational.

Subjects to be covered include:

- Effects of climate on equipment
- Jungle operational techniques
- Preventive maintenance
- Recovery and repair techniques

Staff members and leaders should receive familiarization training on these techniques in order to supervise the operators. In addition, staff and leaders should learn those special supply requirements and procedures in the operational area. They should also be familiar with the capabilities of those logistical units supporting the force.

CHAPTER 4

The Threat in Jungle Areas

Section I. GENERAL

The jungle is an environment which stretches in a broad belt around the tropical areas of the world. Each of these areas has its own military, political, and economic conditions. As a result, it is impossible to describe one threat which applies to all jungle areas. Potential enemies which US forces might face in the jungle run the spectrum from lightly armed guerrillas all the way to conventional forces.

This chapter describes the main features of guerrilla and conventional forces as they are found in the jungle, and briefly outlines the types of potential threat forces in various jungle regions.

Section II. GUERRILLA FORCES.

WHAT GUERRILLAS ARE

Guerrillas are irregular forces. They normally constitute the military faction of a political resistance or a subversive movement. These forces engage in unconventional operations in order to undermine the power of an established government or to take political control away from other factions. Their goal is normally to establish a new government, often according to a radical political philosophy.

The basic guerrilla organization is a three- to five-man cell. These cells are capable of independent action. They also can be brought together for larger operations and dispersed later. Guerrillas are organized into cells for two reasons. One is for security. The fewer the people who can identify members of a guerrilla force, the better the chances are that it will survive. The second reason is for support. Guerrillas must live off the land to a large degree, and small cells are easier to support in this manner.

CONTENTS

HOW GUERRILLAS FIGHT

Guerrillas are usually weaker than conventional forces in terms of total resources. For that reason, guerrillas will not attempt to overwhelm large units of their opponents in combat. They will instead try to inflict as much damage as possible in lightning actions, withdrawing before the opposing forces can react. Guerrillas are most effective when they strike widely separated targets over a long period of time. This type of action will confuse, demoralize, and frustrate their opponents.

Typical missions which guerrillas conduct to accomplish their goals include:

- Destroying or damaging vital installations, equipment, or supplies
- Capturing supplies, equipment, or key governmental or military personnel
- Diverting government forces from other operations
- Creating confusion and weakening government morale

These missions are not normally accomplished by the use of conventional attacks and defenses. Instead, guerrillas rely on speed, surprise, and security. Guerrilla operations include raids, ambushes, mining and boobytrapping, and sniping.

Targets are selected by the guerrilla based on an analysis of how much the elimination of the target will disrupt the government, what the effect on the populace will be, the risk of being killed or captured, and the amount of weapons or supplies which can be seized. This analysis calls for timely intelligence, which is gained by active patrolling.

The retention of the initiative is the key to success in guerrilla operations. Guerrillas rely on their ability to strike where they are least expected, at points where the government forces are least prepared. If the guerrillas lose the initiative, and are forced to react to the operations of conventional forces, their effectiveness is greatly reduced.

Guerrillas are not normally organized or equipped for stand-and-fight type defensive operations. They prefer to defend themselves by moving, by dispersing into small groups, or by diverting the opponent's attention while they withdraw. Whenever possible, these operations are accomplished by offensive operations against the opponent's flank or rear. If the government forces persist in their attack, the guerrillas are prepared to disengage to keep their freedom of action. If forced to disperse into small groups, the guerrilla forces become less effective until they regroup to resume offensive operations.

One of the most important needs of guerrilla forces is support. This support can come from a number of sources. Food, for example, can be stolen or supplied by political sympathizers. Weapons can be gathered from raids on government installations. A foreign power may provide secret training, and shipments of food, weapons, ammunition, and equipment. If the guerrillas can be cut off from these sources of support, they will be much less effective.

To protect their operations, jungle guerrillas will normally establish bases from which they can operate. These bases will be in remote areas. The bases will be secured by a combination of guerrilla outposts and by a grapevine intelligence network established by political sympathizers. Although they may be difficult to find, there will normally be concealed routes into the bases, from which the guerrillas have access to their targets and sources of support.

GUERRILLA STRENGTHS AND WEAKNESSES

Guerrillas operate most effectively in countries where the people are discontented with government policies. If the people are apathetic or passively hostile to their government, the guerrillas will seek to develop this feeling into a popular base of support. If no such feeling exists among the people, it will be much harder for guerrillas to set up operations.

Some common guerrilla strengths include:

- Highly motivated leadership
- Strict, swift discipline among the guerrillas and sympathizers
- Strong belief in a political, religious, or social cause
- Capability to raise or lower the level of intensity from subversion to open warfare

Some common guerrilla weaknesses include:

- Mental and physical stress, caused by long periods of isolation in an unstable environment
- Fear of criminal prosecution by the government, or of reprisals against friends and family
- Feeling of numerical and technological inferiority to counterguerrilla forces
- Uncertain public base of support
- Requirements to secure supply lines, transport means, and storage facilities or caches

Section III. CONVENTIONAL FORCES

TYPES OF CONVENTIONAL FORCES IN THE JUNGLE

Conventional forces committed to jungle operations can perform any one of a number of missions. The lowest level of involvement is the use of conventional forces to advise and assist native guerrilla or paramilitary forces, teaching them either how to fight or how to operate sophisticated equipment. A higher level of involvement is the use of conventional forces as a military cadre in units which are composed of native forces. Finally, the highest level of involvement is the operation of conventional forces in a conventional role, fighting major battles in the jungle.

Conventional jungle enemies may come from a number of places. It is possible that US forces committed to jungle operations will fight native conventional forces. It is also possible that US forces will fight conventional forces brought in from a sponsoring hostile power. In either case, most potential jungle enemies are infantry forces, supported with artillery, mortars, and armored vehicles, organized along the lines of Soviet forces. These forces may also have a capability to conduct tactical air (TACAIR) operations and nuclear, biological, chemical (NBC) warfare. They may be equipped with weapons and equipment that are a generation or two older than those found in more modern armies.

HOW CONVENTIONAL FORCES FIGHT IN THE JUNGLE

The way in which a potential conventional threat army fights in the jungle depends on the terrain, the combat experience of that army, and the degree to which it models itself after the forces of a sponsoring power.

In general, however, jungle enemies can be expected to follow these tactical principles:

- **Maintain the offensive; defend only to gain time.**
- **Embrace the enemy; stay close to reduce the effects of his firepower.**
- **Infiltrate at every opportunity.**
- **Operate during periods of limited visibility.**
- **Use surprise tactics; raids, ambushes, and patrols.**

A jungle enemy can be expected to be skilled in the art of camouflage, the chief means he has to evade his opponent's firepower. Camouflage will be for him a way of life. He will probably use bunkers and tunnels as protective survival measures. To slow opposing forces, he may use obstacles, mines, and boobytraps. He will move on covered and concealed routes, using darkness to conceal most of his operations. He will probably depend heavily on streams and rivers to provide concealed routes of movement and drinking water.

He can be expected to remove all intelligence indicators from the battlefield. He will go to great lengths to remove his dead, wounded, weapons, and even expended cartridges from the battlefield. He will try to leave no information relating to order of battle, strength, dispositions, or intentions.

"We captured numerous enemy documents which either condemned or commended certain units for the police of the battle field"

—Report, 25th Infantry Division, Kontum Province, Republic of Vietnam

Jungle enemies have also used deception means, such as explosive bullets and firecrackers, to mislead US units as to the size and disposition of the forces opposing them. Communications deception and jamming have also been used by jungle enemies against opponents.

Since the US Army is noted for employing an abundance of firepower, jungle enemies in the past have preferred to engage US units at extremely close range. At times, it is impossible for US commanders to use their supporting indirect fires without taking friendly casualties. The specific effects that this technique has on offensive and defensive operations will be discussed later. In general, however, the US ground commander must operate in such a manner that all his fire support can always be used effectively.

Jungle enemies can be expected to train hard to use the jungle to their advantage. In the offense, for example, they use the thick foliage to infiltrate positions and eliminate command posts (CP), key weapons, and vital facilities. These operations are conducted to take away their opponent's advantages in command and control, fire support, and logistical means. The intent is to put their forces on a more equal footing with their opponents. This situation can be exploited by a force with superior knowledge of the terrain.

When forced to defend, these forces will quite often prepare elaborate defensive positions, well camouflaged and concealed. In addition, defenders may use snipers, boobytraps, and ambushes to delay, create a sense of confusion and insecurity, and cause the attacker to surrender the initiative.

It is also possible that US forces committed to jungle operations will fight Warsaw Pact forces, probably members of airborne divisions. Although these troops will have newer and more sophisticated weapons than some of the troops native to jungle areas, *they probably will not be familiar with the local terrain and may not be well trained in jungle operations.*

STRENGTHS AND WEAKNESSES OF CONVENTIONAL JUNGLE ENEMIES

Much of a conventional jungle enemy's effectiveness depends on familiarity with the terrain. In general, this means that armies native to a battlefield area will be more effective than forces from outside. Even if these outside forces have a greater amount of firepower than the native forces, the lack of terrain familiarity may limit their ability to use that firepower.

Weaknesses of potential conventional jungle enemies will probably include:

- Larger units, much more difficult to
- Bigger targets for close-air support or artillery
- More difficulty in evading detection
- Less information from local sympathizers

Strengths of potential conventional jungle enemies will probably include:

- Adequate firepower for conventional attacks and defense
- Knowledge of the terrain and area
- Well-trained and disciplined soldiers
- Independence from local support

HOW THE THREAT DEFENDS

The threat defense is a temporary measure, adopted only when necessary. This does not imply, however, that the threat defense consists of half measures or that he is unskilled in defense techniques. Jungle enemies will use every trick possible to survive against massive amounts of firepower. His defense will be cleverly and carefully prepared.

A typical jungle enemy defensive position consists of a complex series of earth and timber bunkers, spider holes, and tunnels. These are positioned to achieve mutual support. Bunkers are built low to make them more difficult to see and engage by fire. They are well camouflaged—even the fields-of-fire may be cut from the waist down, so that they will be unnoticeable to a standing man. Weapons positions are planned to provide interlocking fires—lethal even during limited visibility. Boobytraps and obstacles are integrated into the defense to slow, demoralize, and confuse the attacker.

The jungle threat's concept of the defense is to trap the attacker by allowing him to move into prepared fields-of-fire. Fire is opened at extremely close range, sometimes at 50 meters or less. This is done for two reasons—first, to bring fires to bear from all sides, and, second, to force the attacker to remove himself before he can call for supporting fires.

While the key part of the jungle threat's defense is automatic weapons positions in bunkers, the enemy will also put snipers in the trees. In this way, the attacker cannot devote his full attention to the bunkers, because he must also deal with the snipers.

If the attacker is too strong, the jungle threat will attempt to withdraw over routes that have been planned and scouted to make the withdrawal as rapid as possible. Stay-behind ambushes, snipers, mines, and obstacles are used to slow the attacker.

HOW THE CONVENTIONAL ENEMY ATTACKS

Attack is the preferred form of combat for potential jungle enemies. Because most of these forces expect to have a disadvantage in firepower and technology when fighting US forces, most of them have developed special techniques to help make up the difference through surprise. They may, for example, probe a defensive position until the defender reveals the location of his key weapons. These weapons are then eliminated by infiltrators before the main attack. They may use firecrackers to create a diversion, drawing the defender's fire and deceiving him as to the size of the attacking force. They may infiltrate the defense to eliminate command posts, mortars, or artillery units.

"Decoy the Americans from one direction by smoke, firing, or shouting. Then attack him from an unexpected direction."

—Captured Japanese Document, World War II

Threat units will avoid attacking prepared defenses when possible. They prefer to attack a weak point, using the jungle, weather, and their own special training as much as possible. Sapper squads are specially trained to infiltrate minefields and obstacles in order to neutralize key positions or create a gap in the defense. The enemy may also isolate a position, so that their opponents will be tied down in trying to relieve it, or they may conduct raids to disrupt operations and lower the defender's morale. Darkness, poor weather, and rough terrain will be used to conceal these operations.

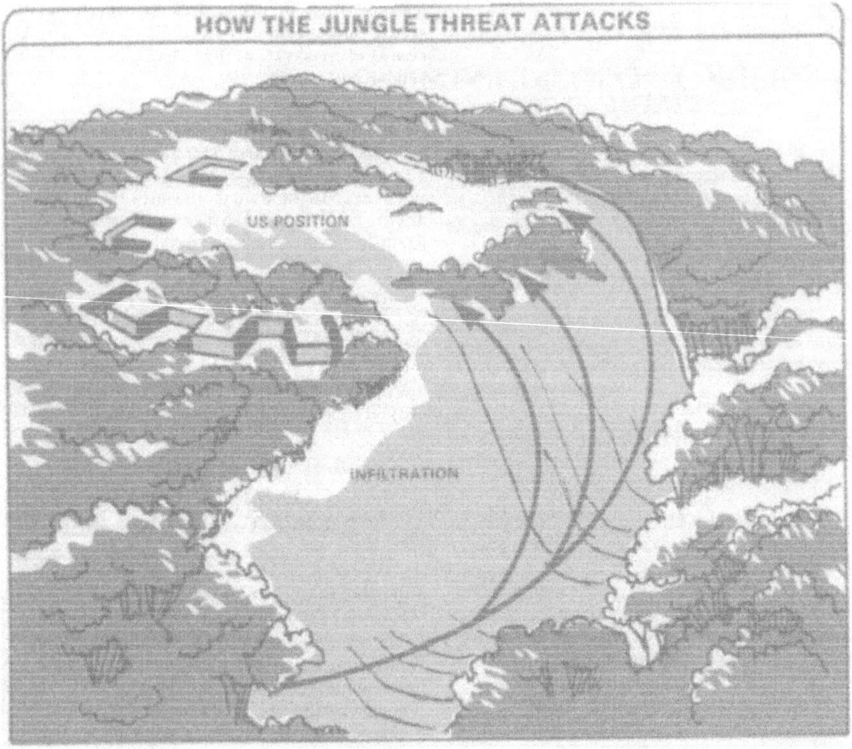

A commander should never assume that any jungle area is impassable to a well-trained jungle enemy. Experience has shown that such enemies are very adept at using extremely difficult terrain effectively as avenues of approach.

"Use fog and rain to catch the Americans off guard. Make an assault suddenly, from positions which the Americans believe unapproachable, such as cliffs, rivers, and jungles."

—Captured Japanese Document, World War II

Although the jungle enemy attacks swiftly, his attacks are planned in minute detail. If he is allowed to attack according to plan, the jungle enemy is an effective force. If the defender can interrupt even a minor part of the plan, the enemy will have difficulty in adjusting, and the attack will probably fail. This aspect of enemy operations places a high premium on the struggle for the initiative at all levels.

If the defending force can be pushed out of its positions and forced to retreat, the jungle enemy will probably make every effort to maintain contact through pursuit. He will harass the rear guard, at the same time sending forces to outrun and cut off the retreating force. He will then try to destroy the retreating force by ambush or encirclement.

Section IV. POTENTIAL ENEMIES IN THREE JUNGLE REGIONS

It is impossible to describe one jungle threat which applies to all areas of the world. There are, however, certain characteristics of potential threat forces that are peculiar to specific jungle regions.

LATIN AMERICA

The most likely threat the US forces may face in Latin American jungles are insurgent movements. These movements aim at the overthrow of a wealthy ruling class to install a new regime. The US is often viewed by the insurgents as an ally of the government, and US facilities and institutions are often targets for these movements.

The military faction of these insurgent movements consists of guerrilla forces similar to those described earlier. They are organized into small cells, are lightly armed, and are capable of concentrating for acts against major facilities and then dispersing after the operation. Although their ultimate objective will often be the establishment of control over the urban areas, they may use the jungle to provide a concealed and secure base of operations. The support of the local people is very important to their survival.

At the time US forces are committed to fight in Latin American areas, guerrilla forces are likely to be augmented with military aid and personnel from other sponsoring countries in the region. These forces may perform any one of a number of roles: advisors, guerrilla cadre, or limited conventional combat. Logistical and intelligence support may also come from these forces.

SUBSAHARAN AFRICA

The conflicts in this region since World War II have been waged by insurgent groups against perceived vestiges of colonialism or imperialism. Most of these colonialist and imperialist institutions are connected in the minds of the insurgents with the Western European powers. As a result, the instability in some areas of this region has provided a tempting target for provocation. To make matters more complicated, many of the conflicting factions are also struggling among themselves, due to political or ancient tribal differences. This in turn creates even more regional turmoil, and an even greater vulnerability for exploitation.

Conflicting factions in Subsaharan Africa consist primarily of guerrilla groups. These guerrillas, however, are often more heavily armed than Latin American guerrillas for two reasons. First, these groups have mortars, artillery, and recoilless weapons from national army formations which have been defeated or disbanded. Second, external powers have backed their favorite factions by supplying arms, ammunition, and equipment. For the most part, these guerrillas subsist by acquiring food and supplies from the countryside.

Foreign involvement in these guerrilla movements has consisted of advisors and cadre from sponsoring nations. Should US forces ever fight in this region, it is likely that they will encounter troops foreign to the nation. In addition, there is also a possibility that Warsaw Pact troops, primarily airborne or tactical aviation units, would be committed to such a region to fight US troops.

SOUTHEAST ASIA

In many respects, the potential threat array in Southeast Asia is the most complicated of any jungle region. There are active guerrilla movements in most Southeast Asian countries as well as tribal and cultural conflicts. There is a good possibility of foreign support or intervention.

The unique development in this region has been the rise of a regional power. Since the end of US involvement in Southeast Asia, this power has developed a potent conventional force, using equipment captured from the US and its allies or supplied by communist countries. More than any other potential threat native in a jungle region, it possesses the ability for sustained conventional operations against any US forces which might be deployed in the area. Its capabilities span the range from clandestine guerrilla operations to large-scale conventional attacks, supported by tanks, motorized units, artillery, and aviation.

Because there are already strong forces in this region, the probability of involvement of large numbers of world power forces is not great. There is a good possibility, however, that US troops committed in these areas might encounter weapons and equipment supplied by a world power. They might also encounter advisors from world powers that instruct and aid the native forces in the use of sophisticated equipment.

Finally, of all the regions discussed thus far, the chemical warfare threat will probably be greatest for US forces conducting operations in Southeast Asia.

Section V. WEAPONS USED BY POTENTIAL JUNGLE ENEMIES

Although potential jungle enemy forces vary widely from region to region, there are certain types of weapons which are commonly found in jungle countries. US forces should become familiar with these basic types of weapons in order to be able to recognize them on the jungle battlefield. They should also have a basic knowledge of the weapons' characteristics and know where the weapons are found in typical communist forces organizations.

Although guerrilla forces do not have the same type of organizational structure as conventional forces, they too will probably carry many of these weapons.

RIFLE PLATOON

The threat rifle platoon normally has three rifle squads of 6 to 10 men each. Weapons found in the platoon will include small arms, one to three light machineguns, and one to three grenade launchers. Typical weapons include:

AKM ASSAULT RIFLE

Recognition features are the pistol grip, the curved-box magazine, and the "underslung" barrel without a bipod. The AKM fires the M-1943 7.62-mm cartridge, also used in the SKS, RPD, and RPK. The AKM also comes with a metal folding stock.

CHARACTERISTICS OF AKM

Caliber and ammo .... 7.62 mm, M-1943

Cyclic rate of fire .... 600 rpm

Practical rate of fire:

- auto .... 100 rpm
- semi .... 40 rpm

Magazine capacity .... 30 rounds

Effective range .... 400 m

AK-74 ASSAULT RIFLE

A new, small caliber version of the tried-and-proven AKM which takes advantage of modern ammunition technology. Two models are being produced: the standard stock AK-74 and the folding stock AKS-74.

CHARACTERISTICS OF AK-74

Caliber .... 5.45 mm

Ammunition .... high-velocity ball (somewhat similar to US M16 rd)

Mode of fire .... automatic and semiautomatic

Cyclic rate of fire .... 650 rpm

Maximum effective range .... 350 m

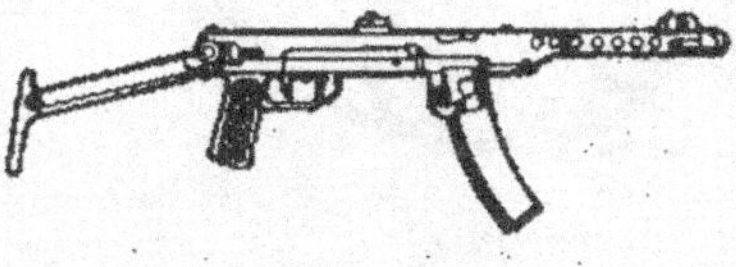

PPS SUBMACHINEGUN

The PPS is a fully automatic weapon. It has a hinged stock which folds up and forward. It also has a compensator welded on front of the barrel jacket. It fires the 7.62-mm M-1930 "P" ammunition from a curved-box magazine.

CHARACTERISTICS OF PPS

Cyclic rate of fire .... 650 rpm

Practical rate of fire .... 100 rpm

Magazine capacity .... 35 rounds

Effective range .... 200 m

RIFLE PLATOON CONTINUED

SKS CARBINE

This light, shoulder-fired weapon is recognized by a characteristic folding bayonet, a sporting rifle appearance, and triangular portion of the magazine which extends through the lower side of the stock, just forward of the trigger guard.

CHARACTERISTICS OF SKS

Caliber and ammo	7.62 mm, M-1943
Operation	gas, semiautomatic
Magazine capacity	10 rounds
Effective range	400 m

M-1890/1930 SNIPER RIFLE

The M-1890/1930 sniper rifle is a Soviet model fielded during World War II. It is still a standard weapon in many satellite armies. The M-1890/1930 mounts a telescopic sight. This sight is similar to the sight used on US hunting rifles.

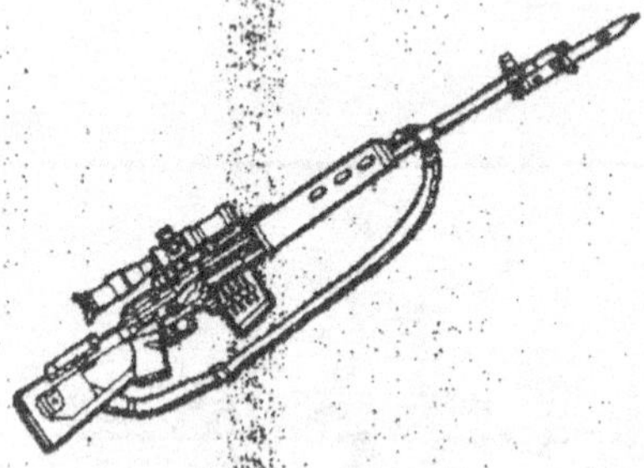

CHARACTERISTICS OF M-1890/1930

Caliber and ammo	7.62 mm, M-1908
Operation	turning bolt
Magazine capacity	5 rounds
Effective range	800 m
Maximum range	1,300 m

RPG-2 GRENADE LAUNCHER

The RPG-2 recoilless antitank rocket is fired from the shoulder. Recognition features are the warhead of the projectile which is larger than the launching tube, and the launcher itself, which is long, slender, and has a pistol grip handle and two sights.

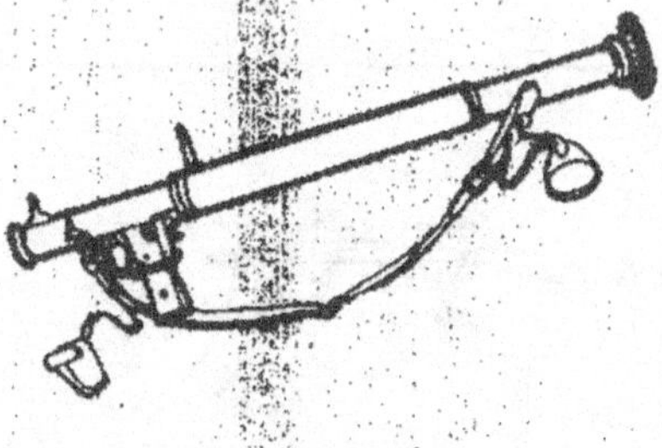

CHARACTERISTICS OF RPG-2

Caliber of tube	40 mm (1.58 in)
Caliber of projectile	82 mm (3.2 in)
Effective range	100 m
Rate of fire	4 to 6 rpm

RIFLE PLATOON CONTINUED

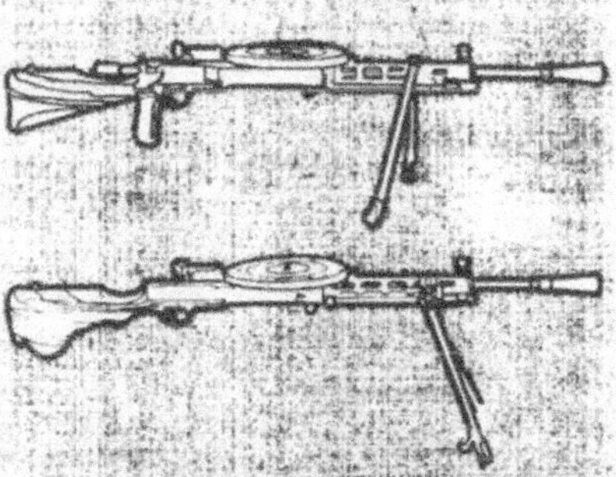

DPM AND DP LIGHT MACHINEGUNS

These machineguns can be recognized by the pan-type drum magazine on top of the receiver, slotted barrel casing, wooden stock with cheek rest, the fixed bipod, and flash suppressor. The DPM has a pistol grip and an operating-rod-spring housing projecting to the rear of the receiver. The DP has no pistol grip, and the operating-rod-spring housing does not project to the rear of the receiver.

CHARACTERISTICS OF DPM/DP LMG

Caliber and ammo	7.62 mm, M-1908
Cyclic rate of fire	550 rpm
Practical rate of fire	80 rpm
Magazine capacity	47 rounds
Effective range	800 m

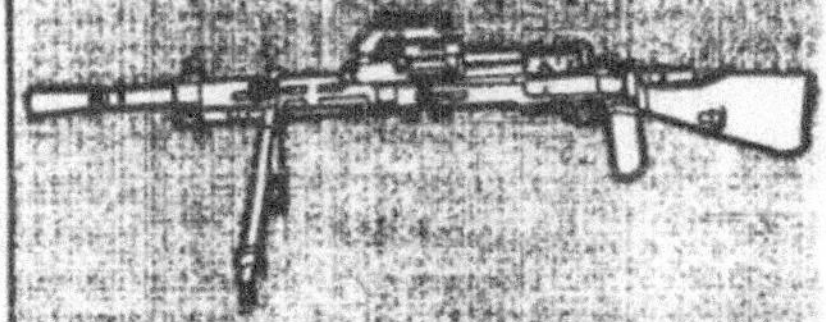

RP-46 LIGHT MACHINEGUN

The RP-46 was developed from the DP series. It has a detachable belt-feed mechanism that can be replaced by a DP type drum-feed mechanism. It is distinguishable from the DP and DPM by its carrying handle and heavier, thicker barrel.

CHARACTERISTICS OF RP-46

Caliber and ammo	7.62 mm, M-1980
Cyclic rate of fire	600 rpm
Practical rate of fire	230 to 250 rpm
Magazine capacity	200- or 250-round belt
Effective range	800 m

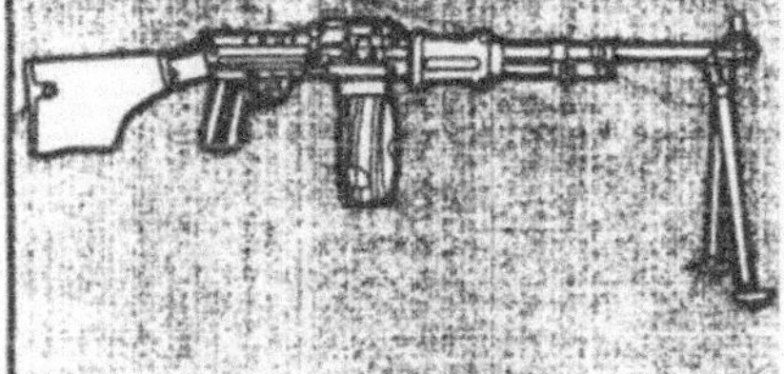

RPD LIGHT MACHINEGUN

The RPD uses the same ammunition as the AK-47 rifle. It has a drum (which houses a metallic link belt), a pistol grip, a permanently attached bipod, and an upper hand-guard which is flared at the ends. The RPD can fire full automatic only and has a chrome-plated barrel.

CHARACTERISTICS OF RPD

Caliber and ammo	7.62 mm, M-1943
Cyclic rate of fire	650 rpm
Practical rate of fire	150 rpm
Magazine capacity	100-round belt in drum
Effective range	800 m

RIFLE COMPANY

The rifle company usually has three rifle platoons and a mortar platoon with two or three 60-mm mortars.

60-MM MORTAR

CHARACTERISTICS OF 60-mm MORTAR

Length of tube	.724 m (28.5 in)
Weight	20.4 kg (45.0 lbs)
Elevation	40 to 85 degrees
Traverse	14 degrees
Maximum range	1,530 m

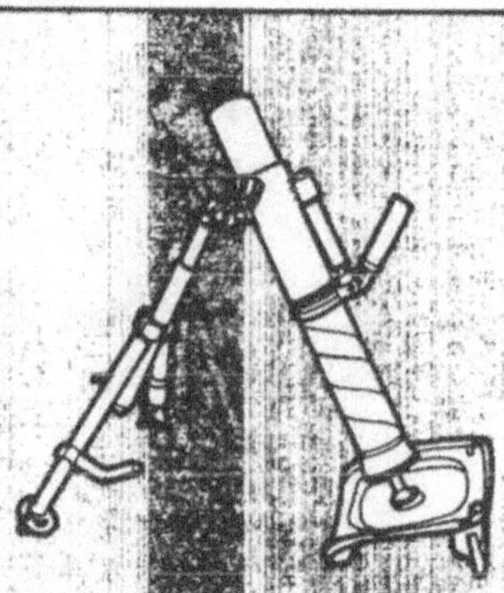

RIFLE BATTALION

The rifle battalion normally has three rifle companies. It may also have a heavy machinegun company, with 6 to 9 heavy machineguns, and a mortar company with 6 to 12 82-mm mortars.

DshK HEAVY MACHINEGUN

The DshK is used in a ground role, and on tanks and assault guns as an antiaircraft weapon. It has radial cooling fins on the barrel. In the ground role it has a wheeled mount and a shield.

CHARACTERISTICS OF DshK

Caliber	12.7 mm
Cyclic rate of fire	540 to 600 rpm
Practical rate of fire	125 rpm
Feed	50-round belt
Effective range; ground targets	1,500 m
Effective range; air targets	1,000 m

RIFLE BATTALION CONTINUED

82-MM MORTAR

There are three versions of the 82-mm mortar, but the M-1937 is by far the most common. All fire the same ammunition and can also use US 81-mm mortar rounds. (The US 81-mm mortar cannot fire the 82-mm ammunition.) The mortar is disassembled into three pack loads for transport.

CHARACTERISTICS OF 82-MM MORTAR

Weight, travel position	56 kg (123 lbs)
Barrel length	1.22m (48.03 in)
Elevation	45 to 85 degrees
Traverse	6 degrees
Maximum range	3,040 m
Rate of fire	15 to 25 rpm
Projectile types	HE and smoke
Projectile weight:	
HE	3.3 kg (7.3 lbs)
Smoke	3.67 kg (8.0 lbs)

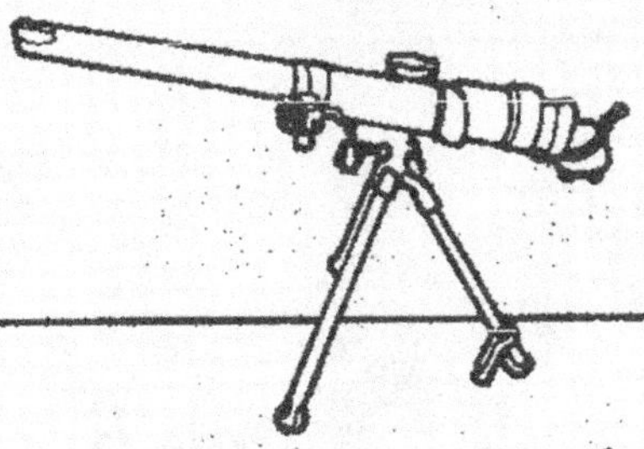

82-MM RECOILLESS RIFLE/TYPE 65

CHARACTERISTICS OF 82-MM RCLR

Weight	28.2 kg (62 lbs)
Caliber	82 mm
Effective range	450 m
Rate of fire	6 rpm
Ammunition	HEAT

75-MM RCLR

The 75-mm RCLR type 52 is a copy of the old US M20. The type 56 is a further development with the same performance as the type 52, but with a different mounting.

CHARACTERISTICS OF 75-MM RCLR

Length of tube:	
type 52	2.260 m (56.92 in)
type 56	2.280 m (58.15 in)
Weight, firing position:	
type 52	85 kg (187 lbs)
type 56	87 kg (190 lbs)
Rate of fire	10 rpm
Maximum range	6,675 m
Effective range	640 m

DIVISION AND HIGHER

Support from division and higher may include fire from 120-mm mortars, 76-mm guns, 122-mm howitzers, or 122-mm rockets.

120-MM MORTAR M-1938/1943

The 120-mm mortar in firing position looks like the 82-mm mortar but is much larger. It can be trigger-fired or drop-fired. An easily attached transport limber is provided for towing the mortar with a truck or armored personnel carrier. In addition to high explosive rounds, this mortar also fires smoke and incendiary rounds.

CHARACTERISTICS OF 120-MM MORTAR

Length of tube	1.854 m (51.66 in)
Weight in firing position	275 kg (605 lbs)
Elevation	65 to 80 degrees
Traverse	8 degrees
Maximum rate of fire	15 rpm
Maximum range	5,700 m
Weight of projectile: HE	15.4 kg (33.9lbs)

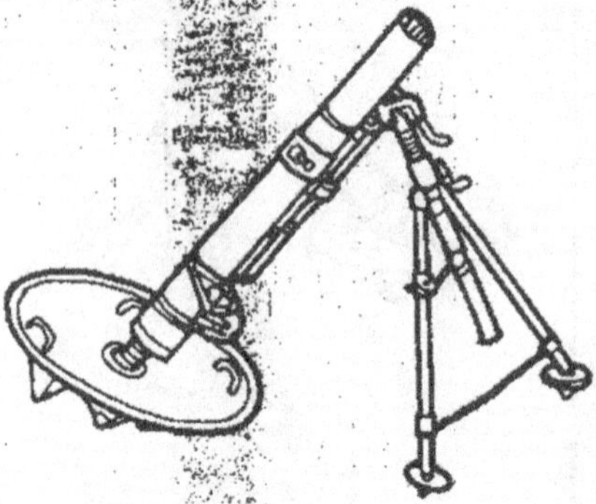

122-MM HOWITZER M-1938 (M-30)

This is a standard divisional artillery piece. The recoil mechanism is in a cradle below the tube, and the recuperator is above the tube.

CHARACTERISTICS OF 122-MM HOWITZER

Length of tube	2.8 m (9.2 ft)
Weight, firing position	2,500 kg (5,500 lbs)
Elevation	-3 to 63.5 degrees
Traverse	49 degrees
Rate of fire	5 to 6 rpm
Maximum range	11,800 m
Weight of projectile: HE	21.8 kg (48.0 lbs)

DIVISION AND HIGHER CONTINUED

122-MM SINGLE TUBE ROCKET LAUNCHER

The Soviet 122-mm rocket can be fired from a single tube launcher, consisting of a shortened BM-21 tube, mounted on a light tripod. The single-fired rocket has less range, but has the same warhead as the version fired from the multiple rocket launcher. The single-fired 122-mm rocket was used extensively in Vietnam; however, currently the BM-21 multiple rocket launcher is the most prevalent.

CHARACTERISTICS OF 122-MM ROCKET LAUNCHER

Length of tube	2.50 m (98 in)
Weight, tube/tripod	21.8/27.7 kg (48.0/60.9 lb)
Height, firing position	1.00 m (39.5 in)
Elevation	-4.5 to 42 degrees
Traverse	14 degrees
Maximum range	11,000 m

THREAT VEHICLES

TYPE 59

The type 59 tank is a Chinese version of the Soviet T-54. The major differences between the type 59 and the T-54 appear to be the type 59's lack of a gun stabilizer and infrared equipment, and the omission of power traverse. Both the gunner and loader have hand-traverse mechanisms. This results in a slow rate of engagement and difficulty in target identification on anything but a flat fire position.

PT-76

The PT-76 is the standard Soviet reconnaissance tank. The suspension of the PT-76 consists of six large, evenly spaced road wheels with no track return rollers. The drive sprocket is at the rear. The turret is circular and has flat sloping sides. The gun tube has a multibaffle or double-baffle muzzle brake. The PT-76 is amphibious. In the water, it is propelled by hydrojets which shoot water through ports in the rear of the tank. The gun fires the same ammunition as the 76-mm divisional gun M-1942 (ZIS-3).

CHAPTER 5

Tactical Operations

Section I. GENERAL

This chapter deals with tactical operations in the jungle. Before reading this material, soldiers should be familiar with the appropriate level of basic tactics in:

FM 7-8, The Infantry Platoon and Squad (Infantry, Airborne, Air Assault, Ranger).

FM 7-10, The Infantry Rifle Company.

FM 7-20, The Infantry Battalion (Infantry, Airborne, Air Assault, Ranger).

FM 90-4, Airmobile Operations.

FM 100-5, Operations.

Although jungle operations are conducted according to the basic guides contained in these manuals, there are special techniques which help to insure success in the jungle. These techniques result from the restricted maneuver, slow tempo, close combat, and limited visibility commonly found in the jungle.

Combat in the jungle is characterized by long periods of developing the situation and looking for the enemy; and short periods of violent, and sometimes unexpected, combat.

To meet these conditions, units must have:

- **Aggressive intelligence-gathering procedures**
- **Disciplined soldiers**
- **Solid SOPS proven in training and updated on a continuous basis**
- **Aggressive and tough-minded leadership**

These four points must be emphasized when a unit is engaged in jungle operations. The need for discipline is evident when one considers the extended periods of looking, often fruitlessly, for the enemy. When contact is made, maximum advantage can only be achieved through aggressive and violent action predicated upon solid SOPs. Aggressive leadership at the small-unit (squad and platoon) level is the one element that ties together the discipline and the training.

CHARACTERISTICS OF THE JUNGLE BATTLEFIELD

The thick foliage and rugged terrain of most jungles limit fields of fire and speed of movement.

CONTENTS

The following limitations may restrict fire and movement:

- Lack of line-of-sight and clearance may prevent visual contact between units, interlocking fires, and the use of tube-launched, optically-tracked, wire-guided missile (TOW) or Dragon missiles.
- Tree limbs may block mortars, flame weapons, 40-mm grenades, and hand grenades.
- Machineguns may not be able to attain grazing fire.
- Adjustment of indirect fire support is difficult due to limited visibility and may have to be accomplished by sound.
- Noise conditions differ in the jungle. There are large numbers of animals in jungle areas, and their noise (or lack of it) can give an indication of something out of the norm.
- Sounds in the jungle do not carry as far as on the conventional battlefield due to the amount of jungle foliage. The result is that noises are closer than first believed.
- Movement through jungle areas is also very difficult because:

  Heat, thick vegetation, and rugged terrain will tire troops rapidly, especially those carrying heavy weapons or radios.

  A lack of roads will hinder resupply and evacuation.

These terrain characteristics make jungle fighting different from fighting on more open terrain. To be effective jungle fighters, soldiers must learn to use these characteristics to their advantage. Potential jungle enemies train to exploit the jungle; so must the US Army.

CHARACTERISTICS OF JUNGLE OPERATIONS

The aspects of terrain and enemy discussed above result in fewer set-piece battles. Rather than conventional attacks conducted against conventional defenses, jungle battles are more often ambushes, raids, and meeting engagements. Battles are not fought for high ground as frequently as conventional battles. *Orientation is on the enemy rather than on the terrain.* Hills in the jungle are often too thickly vegetated to permit observation and fire, and therefore do not always qualify as key terrain. In the jungle, roads, rivers and streams, fording sites, and landing zones are more likely to be key terrain features.

The frequency of ambushes, raids, and meeting engagements makes it very important that units in the jungle practice immediate action drills. In the jungle firefight, the side which initiates contact and gains fire superiority in the first few seconds will normally have a decisive advantage.

CONTROL

Command and control are difficult in the jungle. The thick foliage allows leaders to see and control only a portion of their units.

To cope with this problem, commanders and leaders must:

- **Plan their operations carefully**
- **Issue mission type orders**
- **Insure that each soldier understands his part of the mission**

In addition, the thick jungle foliage and heavy monsoon rains often weaken radio signals, making communications difficult. To reduce the effects of the problem, use of the helicopter as a command and control vehicle is recommended. In that the heavy monsoon rains may not allow helicopters to always fly, an alternate means of command and control must be planned for.

FLEXIBILITY

While an appreciation of battlefield characteristics, jungle enemies, and characteristics of jungle operations is useful, flexibility is important to any leader involved in jungle operations. Successful operations require an extraordinary command adaptability—sometimes, a departure from orthodox thinking in favor of new and often untried procedures. Soldiers must learn to live *with* the jungle and adapt to its initially apparent disadvantages. Having done this, the unit can concentrate on the use of concealment, covered movement, and surprise.

SECURITY AND INTELLIGENCE

Commanders must stress effective security measures and aggressive intelligence-gathering techniques to prevent being surprised. The key is to give the front-line soldier an appreciation of the things to look for. Food remnants and feces can indicate how long ago an enemy unit occupied an area. Captured documents, equipment, and weapons may provide order of battle information and an idea of the enemy's logistical situation. Even an ammunition crate may yield a lot number and packing date. From this an intelligence specialist may be able to trace the enemy unit's place in the order of battle.

In the past, US forces operating in jungle warfare have generally been augmented by native scouts, attached down to platoon level. These scouts were auxiliaries, paid by the unit they supported from a fund established by higher headquarters for that purpose. Scouts familiar with the terrain and the enemy can be an extremely valuable asset. Local security regulations should provide guidance as to what friendly information can be given to scouts.

Surveillance, target acquisition, and night observation (STANO) devices, especially infrared, starlight scopes, and unattended ground sensors, are quite effective in gathering information about troop movements in the jungle. Radars and photography are not as effective because of the concealment of the foliage.

The local populace is one of the most valuable intelligence sources. Whether hostile, friendly, or indifferent, the people can provide information which, when processed, will help complete the intelligence picture.

JUNGLE INTELLIGENCE

Security prevents the enemy from gaining intelligence on US units. Active security measures, such as patrolling and the use of observation posts (OP), helps prevent US units from being ambushed or attacked by surprise. These measures do not lessen the need for passive security. Camouflage and noise and light discipline conceal US forces from enemy observation. To prevent being tracked by the enemy, bivouacs and trails must be policed. Odor discipline is also a security measure. The enemy can follow such odors as heat tabs, cigarette smoke, deodorant, and C rations.

TROOP-LEADING PROCEDURES FOR JUNGLE OPERATIONS

STANDING OPERATING PROCEDURES

A unit's jungle operations SOP should include actions which the unit does on a routine basis or actions that are earned out essentially the same way each time they are done. Examples of such actions include organizing for combat, resupply, bivouac and shelter preparation, movement techniques, and battle drill. Use of SOPs will save planning time.

PLANNING USE OF TIME

A unit planning for jungle combat follows the same planning sequence as in any other type of combat operation. In planning the use of available time, leaders must consider that many tasks in the jungle take more time than the same tasks in other environments. More time must be allowed for movement and security. This means that units may have to begin movements earlier in order to accomplish their missions within a specified time. This may leave less time for planning and preparation.

AIRMOVEMENT PREPARATION

If the unit is to be inserted by helicopter, the commander should go through the preparatory steps outlined in *FM 90-4*.

INSPECTION

Prior to beginning a misssion, unit leaders should inspect their troops to insure that:

- **They have all their needed equipment**
- **They have no unnecessary equipment**
- **Weapons are cleaned, lubricated, and zeroed**
- **Equipment and weapons are in working order**
- **Everyone understands his job and the unit's SOP**

CONTROL OF EQUIPMENT

Those troops who carry extra equipment should not be allowed to discard it. Captured US equipment has been used by jungle enemies in the past. In Southeast Asia and in the Pacific, recovered equipment was a major source of enemy supply.

SUPERVISION

Supervision must continue throughout the conduct of the operation. As the troops become tired after long periods of marching or digging in, they will tend to get lax and ignore good security habits. This is an especially common trend if they have not been in contact for a few days. Tight supervision is a must to insure that security patrols and OPs are dispatched and doing their jobs; that troops remain alert; and that fire, noise, and light discipline are not relaxed.

JUNGLE MOVEMENT

PLANNING AND ROUTE SELECTION

Before conducting a move in the jungle, leaders should make a map and aerial photograph reconnaissance. This reconnaissance will indicate possible danger areas, obstacles, and roads or clearings suitable for resupply.

In planning the route, leaders should consider the following:

- **Lines of drift, such as ridgelines, are easy to guide on because they avoid streams and gullies and because they are usually less vegetated.**
- **Danger areas, such as streambeds and draws, are usually more thickly vegetated. They offer excellent concealment, but travel along them is slow and difficult.**
- **Roads and trails should be avoided. Although they are easy to move on, they offer little concealment. These are the areas most likely to be under enemy observation. They are easy to ambush and are very likely to be mined or boobytrapped.**

MOVEMENT TECHNIQUES

Units moving in the jungle should normally use the jungle movement technique, but may use traveling overwatch and bounding overwatch when necessary. The file formation should be avoided in all but the most thickly vegetated areas.

To effectively use the jungle movement technique, certain key factors must be understood. They include the following:

- **Only the platoon should employ this movement technique.**
- **The lead fire team of the lead squad is always in a wedge (modified).**
- **The support elements may move with the headquarters element or be attached to a squad(s) depending upon likely threats.**
- **Each squad maintains an azimuth and pace.**
- **Immediate action drill (SOP) is essential.**
- **This technique is most effective during daylight movement.**

This movement technique is basically characterized as a formation of multiple columns which are mutually supporting.

Advantages of the jungle movement technique:

- Centralized control.
- Rapid deployment to maneuver or reinforce.
- Ease of movement (three routes).
- 360-degree security during movement and at halts.
- Multiple navigational aids (three azimuths and pace counts).
- Flexibility of adjustment during movement (danger areas, choke points).

Disadvantages of the jungle movement technique:

- Possibility of loss of contact at major obstacles due to multiple routes.
- Vulnerability to effectiveness of indirect fire weapons.

JUNGLE MOVEMENT TECHNIQUE FORMATION

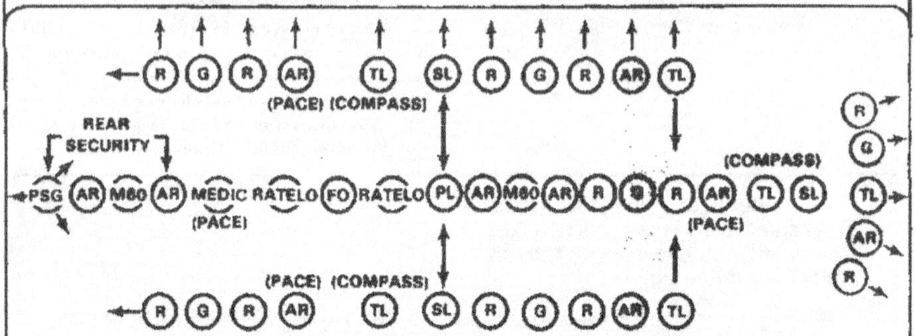

Note the positions of the platoon leader and adjacent squad leaders during movement. It is easily visualized how this formation can "balloon" its flanks at halts to form a hasty perimeter. Also, note the locations of the machinegun teams that allow for rapid attachment. The formation shown depicts a 45-man platoon; however, it can be effectively used with lesser numbers.

In traveling overwatch, the lead element performs the mission of point security, with troops from the rest of the unit performing rear and flank security. When contact is imminent, the unit moves into bounding overwatch. Bounds, as terrain allows, are normally 50 meters or less.

SECURITY

The thick foliage makes ambush a constant danger. Point, flank, and rear security teams will help keep a force from being ambushed. These teams must be far enough away from the main body that if they make contact the whole force will not be engaged. They should not be so far away, however, that they cannot be supported. These security elements must be alert to signs of the enemy, and should carry as light a load as possible so they are able to maneuver. Security duties should be rotated often to avoid fatigue. Scout dogs may also be used with the security element. These dogs often detect the enemy before he is detected by humans. The jungle heat is hard on them, however, and they must be rested frequently.

If contact is broken between elements or individuals, the rear element should remain in position. Those in front should return to establish contact.

HALTS

Units should plan halts on terrain which lends itself to all-round defense. During short halts, soldiers drop to one knee and face outward, their weapons at the ready. If the halt occurs at a trail crossing, security elements are sent out along the trail. The security element remains in place until the unit clears the crossing. During longer halts, units establish a perimeter defense. They run security patrols around their positions, and employ Claymore mines and early warning devices. Before an overnight halt, units should stop while there is still enough daylight to establish a secure perimeter defense, prepare ambushes, and dispatch patrols as necessary. If halted units are separated, connecting patrols should be run periodically to detect enemy infiltration.

Section II. RECONNAISSANCE, SURVEILLANCE, AND SECURITY OPERATIONS

RECONNAISSANCE

Reconnaissance operations are always important in jungle warfare. Many offensive operations in the jungle take on the aspects of a reconnaissance operation during their early stages. This is because the success of offense in the jungle depends on ability to find the enemy. The excellent concealment found in the jungle enables the enemy to operate unobserved both by day and night. Extensive patrolling is necessary to obtain information on his locations, strength, and disposition.

Reconnaissance is a responsibility of all leaders during jungle operations. Units with the capability to conduct reconnaissance should conduct frequent short patrols during the conduct of normal missions. These patrols should be coordinated with higher and adjacent units. In some situations, reconnaissance may become the primary objective of a major jungle operation.

Planning for a jungle reconnaissance should be thorough and well coordinated. Coordination with higher and adjacent headquarters will help insure maximum results from each patrol and eliminate duplication of effort. Radio is the primary means used to control reconnaissance operations. Each reconnaissance patrol must be prepared to make contact, develop the situation, and report to its controlling headquarters.

GROUND RECONNAISSANCE

In the jungle, these operations are accomplished by means of OPs and long- or short-range reconnaissance patrols. The value of OPs is somewhat reduced in the jungle because of the limited visibility. OPs are most effective when used along trails, roads, and streams to detect enemy movement. (Although OPs in other areas may not provide much useful information on the enemy, they still are effective in providing early warning.) Reconnaissance patrols in the jungle are normally squad-size. These patrols move in a manner to take advantage of natural concealment, and avoid becoming engaged with the enemy. The use of helicopters increases the depth behind enemy lines that such patrols can be employed. For further details on reconnaissance patrols, *see section VI.*

VISUAL RECONNAISSANCE

From the air, key terrain features can often be identified and the enemy detected in areas where there are gaps in the jungle canopy. Aerial photographs are important sources of information because photograph interpretation can disclose hidden enemy camps not visible to the air observer. Photographs can also be used to locate helicopter landing zones. Decoy reconnaissance flights can be used to confuse or deceive the enemy about upcoming operations.

RECONNAISSANCE-IN-FORCE

This method may be used to gather intelligence which cannot be gained by any other reconnaissance means. Examples of such intelligence include the enemy commander's plan for committing his reserves, or the trails used by a guerrilla force.

The commander conducting the reconnaissance-in-force will normally organize his unit into a number of reconnoitering forces. These reconnoitering forces conduct movements to contact, hasty or deliberate attacks, raids, reconnaissance, or patrols. Once the unit makes contact with the enemy, the commander must react on the intelligence gained. He must be prepared to exploit success or, if necessary, extricate the forces.

The size of the unit that conducts a reconnaissance-in-force depends on the nature of the intelligence to be gained and the chance that the reconnoitering force will have to fight on unfavorable terms. For example, if a battalion commander wants to find out how an enemy commander will commit his reserve, he may conduct a reconnaissance-in-force with his companies conducting limited objective attacks. If, on the other hand, a commander wants to find the routes used by a number of small guerrilla groups, the reconnaissance-in-force mission may be assigned to a company, which in turn will have its platoons conduct movements to contact.

RECONNAISSANCE-BY-FIRE

When using this method, the force fires on suspected enemy positions to cause the enemy to disclose his position by moving or returning fire. Reconnaissance-by-fire risks the loss of surprise. Its most effective use in the jungle is to find the flanks or gaps in enemy lines. Reconnaissance-by-fire from attack helicopters will often reveal the location of well-concealed enemy troops. Likewise, a reconnaissance-by-fire from armored vehicles firing into a wood line, either while moving crosscountry or along a road, can neutralize an enemy ambush. When using reconnaissance-by-fire, commanders must consider the difficulties of ammunition resupply in the jungle.

SURVEILLANCE

Surveillance operations in the jungle include using all techniques for establishing a continuous, thorough watch of the battlefield. This watch must be established both over large jungle areas and at selected key points such as trails, streams, and clearings. Surveillance operations are usually planned to support other missions.

GROUND SURVEILLANCE RADARS

These radars are best employed in those jungle areas where vegetation and terrain do not restrict line of sight. Night observation devices are also useful in such areas during periods of darkness. Unattended ground sensors, which are not affected by poor line of sight, are very useful in watching specific key areas. These electronic devices are affected by poor weather and are difficult to move in thickly forested areas. As a result, the use of manned OPs in jungle areas should always be planned.

AIR SURVEILLANCE

Surveillance of jungle areas from the air is most effective when pilots are familiar with ground operations and can recognize changes from normal patterns. Repeated flights by the same crews will attain this level of familiarity. In addition to visual surveillance, photographic coverage of an area can assist the surveillance effort. Side-looking airborne radar (SLAR) is not very effective in thick foliage, but can be used for surveillance along roads, trails, or streams. Likewise, infrared detection devices are limited by fog, clouds, rain, and vegetation. Airborne personnel detector devices (sniffers) were developed during the Vietnam war to detect human odors. These devices are extremely effective in detecting base camps of nonmechanized forces, but are limited by fog, rain, and windy conditions.

SECURITY OPERATIONS

Security must be a primary part of all jungle operations; therefore, specific security measures are covered as they apply to other operations throughout this chapter. Compared with operations in other types of terrain, security measures in the jungle must be intensified because of the poor observation and difficulties of control and movement. Operations must be slower than normal, and security forces must be closer to the units secured in order to provide adequate security in the jungle. Because it provides all-round security, the perimeter defense will be the defensive technique used most often by units operating independently in the jungle.

The "stand-to" is an important security technique in jungle fighting. When a unit stands-to, all of its soldiers don their fighting loads and occupy their fighting positions. The unit is 100 percent alert and ready to fight an attacking enemy. Stand-to procedures differ from unit to unit, but common stand-to times are before first light, before last light, before helicopter resupply, and before movement.

Section III. OFFENSIVE OPERATIONS.

SPECIAL FACTORS

The purpose and fundamentals of the offense as outlined in field manuals for other environments generally apply as well to offensive operations in the jungle.

There are, however, factors which require the use of special offensive techniques:

- **Thick foliage makes it difficult for leaders to control their soldiers or to detect the enemy**
- **Fire support is difficult to observe and adjust**
- **Momentum and speed are difficult to maintain**

FUNDAMENTALS OF THE JUNGLE OFFENSE

When considering the use of special offensive techniques, commanders must remember that some offensive fundamentals acquire a new significance in the jungle.

Probably the most important and most difficult of these fundamentals is the requirement to *see the battlefield.* Above all else, the attacker must know the battlefield. As a result, he relies heavily on security patrols, information provided by air and ground reconnaissance, and proper movement techniques.

Key to effective operations in jungle warfare is the fundamental of using weapon systems to their best advantage. In addition to organic weapons, the ground commander must closely coordinate the employment of the supporting weapons available to him. The dense foliage found in some jungles may prevent heavy weapons from moving directly with the infantry. In those cases, TACAIR support and helicopter weapons must make up the difference. TOWs and Dragons, on the other hand, are of limited use in most jungle environments. The soldiers that man these weapons may be more effective as security forces or as reinforcements for maneuver elements. The primary jungle weapons are individual infantry small arms, supported by machineguns and mortars.

To *concentrate overwhelming combat power* against enemy weakness in jungle operations, the attacker must be able to bring up other elements quickly to support an element that is engaged. In determining how far he can separate his subordinate units, the commander must consider the factors of mission, enemy, terrain and weather, and troops and time available (METT). The separation may be expressed in time or in distance. The ability to provide mutual support must not be overlooked. Gaps between units should be covered by scouts and connecting patrols. Targets should be planned along the unit's route so that supporting fires can be responsive. The use of helicopters will permit even more rapid concentration of forces and provide additional firepower *(chap 6).*

The jungle also increases the difficulty of efforts to *provide continuous support.* The key is constant planning, coordination, and maximum use of helicopters. Fires must be planned along the attack route so that they can be delivered in the shortest amount of time. Procedures for calling attack helicopters must be standardized and rehearsed. Likewise, combat service support must be timely and responsive. Ammunition and water must be loaded on pallets in the trains so that they can be brought forward as soon as needed.

CONDUCT OF THE JUNGLE OFFENSE

Because it is so hard to gather intelligence, jungle offensive tactics must be characterized by continuous reconnaissance. A unit attacking without timely information on the location of the enemy may subject its elements to enemy ambush without being able to support them. In such a situation, they may be defeated in detail.

"It is a situation that too frequently occurs in the Vietnam fighting. The forward element, losing men and becoming pinned down, compromises the position of all others. What has started out as an attack loses all form and deteriorates into a costly rescue act."

-S.L.A. Marshall, BIRD

Successful jungle attacks usually combine dispersion and concentration. For example, a rifle company may move out in a dispersed formation so that it can find the enemy. Once contact is made, its platoons close on the enemy from all directions. In this way, they move to support each other and destroy the enemy.

Operations are enemy-oriented, not terrain-oriented. Wherever the enemy is found, that is where he should be destroyed. If he is allowed to escape, he will only have to be found again, with all the risks involved.

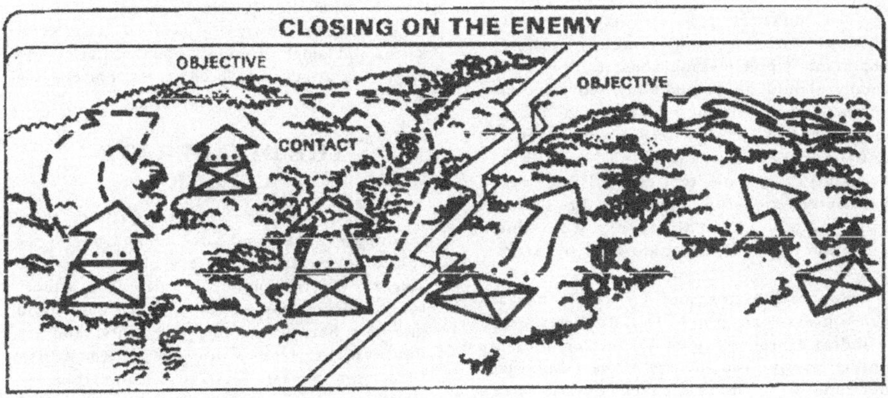

MOVEMENT TO CONTACT AND HASTY ATTACK

These two forms of combat are discussed together since they account for the majority of combat actions in the jungle. The successful follow-on action to movement to contact is a violently executed hasty attack. This action capitalizes on the advantage of surprise and the confusion prevalent in the jungle. Movement to contact in the jungle differs little from the general concept described in FM 7-20. The decision to employ single or multiple columns may depend solely on terrain and foliage considerations.

For companies and battalions, multiple columns are a sound movement practice because this formation provides more firepower to the front and because it is easier to deploy troops from two or three columns than from one file. In order to speed up deployment, units should develop and rehearse SOPs and immediate action drills. Troops should try to maintain a distance of five to seven paces between men, but must maintain visual contact. All-round defense and security measures must be maintained throughout movement.

Once contact with the enemy is made, the unit's first action is to build up a large volume of fire. The commander then assesses the situation and deploys his unit to overrun the enemy's positions while they are still suppressed. In this way, he seizes the initiative. There should be no delay in the troops' movement from the march formation into assault formation. Security elements protect the rear and prevent the enemy's counterattack. These forces may be used later to exploit a success, but should not be committed until the commander understands the situation.

The slowness of jungle maneuver makes a rapid call for supporting indirect fire important. Upon making contact, fires on the enemy should be immediately requested and adjusted from planned targets. To receive effective and timely fire support, accurate and continuous land navigation is necessary. Means for controlling attack helicopters are also important; this includes both radio and visual means. Adjustment techniques should be established by SOP.

Supporting fires and TACALR or attack helicopters can place fires on suspected withdrawal routes, placing further pressure on the enemy. The success of the hasty attack depends to a large degree on the unit's vigorous execution of unit SOPs and the leadership of the squad and platoon leaders.

As the situation is developed and an enemy position is located, a violent assault should be made over the enemy's position. Soldiers stay on the alert for hidden enemy positions, snipers in the trees, and tunnels through which the enemy might move to attack the attacker's rear. This thorough technique will also provide enough information of the enemy and security to permit the commander to use his reserve force for exploitation, if needed.

The assault should be made using fire and maneuver. Soldiers should cover each other, moving by crawls and short rushes. Fire should be well-aimed shots and short bursts of automatic fire.

In such a fast-moving situation, it might be possible, for example, for a platoon to receive a fragmentary order (FRAGO), move to a pickup zone (PZ), and conduct an air assault to an objective. In this case, the planning might take place on the PZ or even in the vicinity of the objective rally point (ORP).

After the objective is seized, it must be secured immediately with a hasty perimeter, OPs, and early warning devices. This is to detect and repel an enemy counterattack or to allow the attacker to prepare to continue the attack.

THE DELIBERATE ATTACK

Based on information gained from reconnaissance and other sources, the commander may formulate a plan to attack a larger objective using a deliberate attack. Jungle terrain favors reduced distances and intervals between troops and units, and the deployment from movement to attack formation as far forward as possible. *In thickly vegetated terrain, the use of some of the same control techniques used in a night attack may be required.*

Fire support is as essential in the jungle as in other types of terrain. Unsupported troops are likely to incur heavy casualties when attacking jungle positions, especially considering the difficulties of employing organic weapons. Targets must be pinpointed by reconnaissance, and fires must be adjusted within very close range of attacking troops. During the assault, these supporting fires must continue until shifted by the assaulting commander. They are then adjusted onto targets which will assist the progress of attacking forces by blocking enemy counterattacks or withdrawal. Due to poor observation, indirect fire may have to be adjusted by sound *(see app I)*.

Assaulting troops move over the objective using aggressive fire and movement to overcome enemy resistance. Assaulting troops again must be alert to snipers, mines and boobytraps, hidden positions, and tunnels which would permit the enemy to maneuver into the rear of attacking forces. Assaulting platoons and squads move in a single direction, with fires concentrated on enemy positions as they are located. Attacking elements must adjust their progress using base elements and phase lines. Smoke may be used to screen the flanks of the penetration from enemy observation and reduce his ability to deliver effective fires. Once an initial penetration is secured, it is exploited until the objective is taken.

After the objective is overrun, it must be secured immediately with a hasty perimeter, OPs, and early warning devices to detect and repel an enemy counterattack or to allow the attacker to prepare to continue the attack *(see IV)*.

INFILTRATION

Jungle areas are ideal for infiltration. Dense vegetation and rugged terrain limit the enemy's ability to detect movement. As a technique to move through the enemy's positions, infiltration can be used with other offensive maneuvers to gain an advantage in the jungle. Although jungle infiltrations are normally conducted on foot, under certain circumstances helicopters or watercraft may be used.

Infiltrations are normally difficult to control. Chances for success are better if troops are well trained, well briefed, and well rehearsed. Roads, trails, and streams should be avoided because they will normally be under enemy surveillance. Movement by stealth is normally slow and exhausting. Phase lines (PL), infiltration routes, and adequate communications must be used to control the operation and to coordinate fires with movement.

CONTROL OF AN INFILTRATION

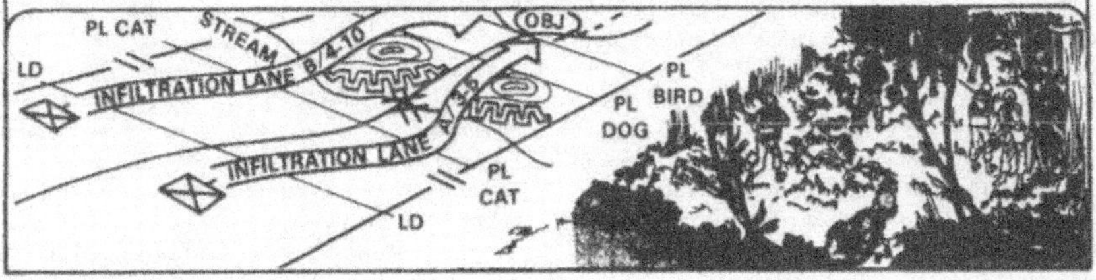

EXPLOITATION AND PURSUIT

Local successes should be exploited as soon as possible to cut off the retreat of isolated enemy forces. Airmobile troops are most effective to block enemy retreat in the jungle. They can also disrupt and harass enemy reserve, logistical, and command operations. Artillery, TACAIR support, and attack helicopters may also be used to block escape routes. Tanks may be used along trails or roads or in less dense areas if properly secured with infantry. During exploitations, rear areas must be secured against the actions of bypassed or infiltrating enemy.

Pursuit operations in the jungle should be conducted to maintain contact with the enemy. The precautions required to secure against ambush can slow pursuit operations considerably. Because attacking troops become more fatigued than defending troops, pursuit operations should be conducted using troops from the reserve. As is the case with most jungle offensive operations, airmobile forces, air cavalry, and attack helicopters can be used very effectively in pursuit operations.

ATTACK AGAINST A FORTIFIED POSITION

As mentioned earlier, jungle enemies have often used strongly fortified defensive positions to protect themselves from the effects of US firepower. If it is necessary to attack such a position, troops will probably encounter bunkers, barbed wire, mines, and boobytraps. The enemy will often have to be burned or blasted out of such positions. These operations will require attacks on a narrow front, great amounts of firepower, and limited objective attacks.

Fortified enemy positions in dense jungle are often so well concealed that troops are not aware of their presence until they have physically encountered them. In these cases, the best course of action is usually to adjust forward dispositions enough to allow use of supporting indirect fires, to deploy additional forces to block possible withdrawal routes, and then to maneuver under the cover of supporting fires to defeat the fortifications in detail.

Maximum use of combat intelligence is required when attacking fortified areas. Aerial photographs, electronic intelligence, interrogation of prisoners of war (PW), and aggressive patrolling are all means of gaining the required information. Patrols also keep the enemy off balance and limit the enemy's ability to patrol.

In addition, combat engineer, and special weapons (such as flame) and equipment should be attached to the maneuver forces to assist in destroying the fortifications.

Destruction of the enemy in those types of positions takes a lot of time and effort. Other means should be used as much as possible. Here is a technique that proved successful in the past:

"The use of CS riot control gas could not be overlooked... On one occasion, a battalion made a night attack with gas masks following an aerial CS attack. A helicopter made several low passes on the windward side of the area and dispersed about 250 CS grenades. This was followed by 20 minutes of artillery fire, about half of which was VT fuze fired into the enemy positions. Behind a walking barrage of artillery fire, one company assaulted, and, once inside the objective, flareships lit up the area. Eighteen enemy were killed, while no casualties were suffered by the friendly troops."

—Report, 25th Infantry Division, Duang Nhgai Province, Republic of Vietnam

Section IV. DEFENSIVE OPERATIONS

SPECIAL FACTORS

The purpose and fundamentals of the defense as outlined in field manuals for other environments also apply to defensive operations in the jungle.

There are, however, certain factors which require the use of special techniques:

- **Thick foliage makes it difficult to detect the approach of an attacking enemy**
- **Slowness of jungle movement makes it difficult to react to an enemy threat**
- **Limited visibility between defensive positions**
- **Limited fields of fire**
- **Psychological impact of fighting in a strange environment**

FUNDAMENTALS OF THE JUNGLE DEFENSE

As in the offense, jungle defensive operations are based on the same fundamentals used in other area operations. Some of these fundamentals acquire a special significance in the jungle.

To succeed in the jungle defense, a commander must *understand the enemy and see the battlefield.* The enemy will probably be expert in using the environment to his advantage, and the defender must understand enemy techniques. To counter the threat of infiltration, the defender must employ all-round defense and all surveillance means available. No amount of electronic means can eliminate the need for frequent patrolling.

The defender must *exploit every advantage* that he has, particularly the abundant concealment provided by the foliage and the weather. A force which remains concealed may disrupt an enemy's attack by using surprise fire from hidden locations. Though not as common as in other types of terrain, features which lend themselves to the defense, such as rivers, gorges, and ridges, should be used if they dominate likely avenues of approach. The defender must appreciate the defensive characteristics of the terrain and environment.

The defender must *maximize the effectiveness of key weapons,* which in the jungle are infantry small arms, mortars, and artillery. The poor trafficability also increases the importance of attack helicopters and TACAIR.

In planning to *concentrate combat power at critical times and places,* the defender must first plan the massing of small-arms fire. Since fields-of-fire will be limited, positions must be placed close together for mutual support. To move troops and weapons rapidly to supplementary or alternate positions, it maybe necessary to cut paths through the bush. Units should rehearse these maneuvers.

As in the offense, the problem of *providing continuous support* is to a large degree solved by effective communications and the use of helicopters.

PLAN OF THE JUNGLE DEFENSE

Planning for the jungle defense should provide for a covering force area, a main battle area, and a rear area. Forces in each area must be provided fire support. Obstacles are planned to improve the natural defensive strength of the terrain. Plans are also formulated for counterattacks. The following factors should be considered when planning for the jungle defense:

Day and night in jungle regions are each roughly 12 hours long. Nights, especially under jungle canopies, are extremely dark. Defensive preparations should begin at least 2 hours before nightfall.

The heat and humidity will fatigue troops rapidly.

Tropical rain will flood positions unless they are adequately drained. During the rainy season, defensive positions should be dug on high ground, if possible.

Because jungle terrain favors infiltration, the use of starlight scopes, OPs, and early warning devices is very important.

After the commander organizes the ground and secures the area to be defended, he then positions his Dragons, if they can be used effectively, and machineguns and clears fields of fire. Leaders must insure that troops do not cut too much vegetation. In order to be mutually supporting, positions will be closer together than on other types of terrain. In addition to attaining mutual support, this helps prevent enemy infiltration. Mines and obstacles should be emplaced where they are covered by friendly fires. These should be located beyond hand grenade throwing range of the defensive positions.

PLANNING THE DEFENSE IN JUNGLE TERRAIN

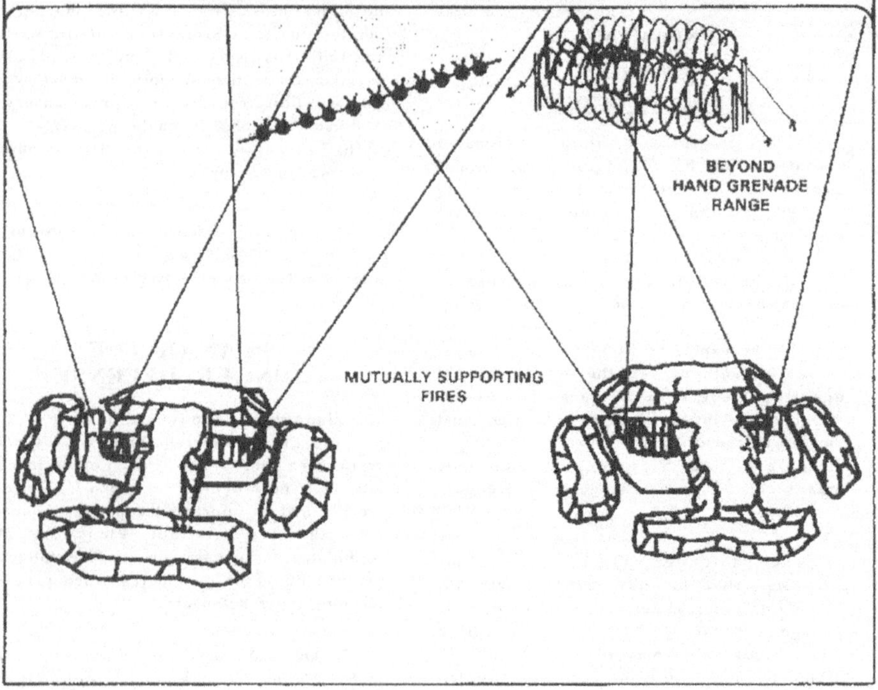

SECURITY

Since enemy tactics, jungle terrain, and bad weather favor attacks conducted by stealth, security should be the leader's first concern. OPs, early warning devices, ambushes, and patrols are all measures which will prevent a unit from being surprised. Patrols must be planned according to an irregular schedule. A system of recognition signals must be used to prevent the engagement of friendly units.

NOTE: If mechanical ambushes are used, units should stop patrolling in that area, and should provide some means, such as communications wire, to guide OP personnel and prevent them from straying into the mechanical ambush.

DEFENSIVE FORMATIONS

The basic defensive technique in the jungle is the perimeter defense. Two other very effective defensive techniques for jungle operations are the triangle and "Y" formations *(see app J)*. Whether a unit is operating independently or as part of a larger defensive position, it must be prepared to defend itself against an attack from any direction. Initially, these formations will be formed by platoons or by companies. Larger units should position their companies in depth to provide all-round defense. Alert and aggressive patrols and OPs will defeat enemy attempts to infiltrate between positions. Later, if time permits, platoon and company formations can be connected with fighting positions and trenches. Even then, however, companies and platoons must be prepared for all-round defense.

PRIORITY OF WORK

Units in the defense must pay particular attention to their priority of work. Since more security measures must be taken than normal, fewer troops will be available to prepare defensive positions at any one time. Positions should be prepared and camouflaged as in any other situation. Overhead cover should be prepared using strong wood and sandbags. Claymore mines and trip flares should be emplaced in front of the defensive positions. Fields of fire should be cut low, leaving enough foliage so as not to reveal the location of the defensive position *(app E)*.

To counter enemy reconnaissance efforts, units should shift the positions of machineguns after dark. After a few days, the entire unit's position should be changed. If a unit remains in position for a longer period of time, it is more likely to be reconnoitered by the enemy and subject to attack.

COMMAND AND CONTROL

Command and control are extremely difficult in the jungle defense. Commanders must place great emphasis on planning, coordination, and small-unit leadership. Decentralized control is important to insure that subordinate units can react to multiple threats. Aggressive leadership at the small-unit level is necessary in fighting off isolated assaults at close range. Alternate communications means must be established wherever possible. An example might be a communications system using wire as the primary means, radio as the secondary means, and pyrotechnics for certain prearranged signals.

Defensive targets for artillery and mortars should be planned on stream and trail junctions, and any other likely enemy avenues of approach. Artillery and mortar fire should also be used to cover the many areas of dead space found in jungle terrain. It is also a good idea to confirm the location of the defense on the ground by using artillery marking missions integrated into registration missions. Signals for the employment of TACAIR and attack helicopters, both day and night, and for medevac and resupply helicopters must also be planned.

CONDUCT OF THE JUNGLE DEFENSE

There are occasions when a unit will have to establish a defense with minimum planning time. This normally occurs when an attack is stalled, at dusk when the unit is still in contact with the enemy and no night attack is planned, or when an intermediate objective must be secured before continuing the attack. These situations are more dangerous in the jungle than in other areas because of the dense foliage and the closeness of the enemy.

The normal course of action in these cases is to establish a perimeter defense. Dragons, if they can be used effectively, and machineguns are positioned immediately where they have the best fields of fire. As soon as possible, OPs and other local security measures are established. Frontages are smaller than in other types of terrain, especially at night, to guard against enemy infiltration. Indirect fires are registered and fighting positions are dug as soon as possible.

Once these actions have been completed, steps are taken to improve the defense. A primary consideration in improving a defense is to expand the perimeter to gain "working room. " This may require limited attacks, massed artillery and mortar fire, or close-in machinegun fire to force the enemy to withdraw. If the position will be occupied for a long time, it should be made as strong as possible.

This may be done as follows:

- **A small reserve is formed as soon as possible to react to enemy threats**
- **Local security is pushed forward**
- **Counterattack plans are developed**
- **Wire communications are established and pyrotechnic signals planned**
- **Machineguns are employed singly in order to cover as many enemy approaches as possible**

If troops remain in a defensive position for a long time, they must not become complacent. Leaders must inspect weapons, positions, and the cleanliness of troops. They must also develop plans for alerts, feeding, maintenance, and bathing. These activities must be scheduled according to a random pattern so that the enemy cannot take advantage of a set routine.

An alert system must be established, so that a portion of the defensive force is always awake. Although the poor observation in jungles favors the enemy's attack at any time, the early hours of the morning afford him the greatest chance of surprise unless positive alert measures are taken. All troops should stand-to before dawn.

The jungle enemy will try to probe a position to locate the flanks of positions and key weapons. Soldiers must not give away their positions by premature firing. Claymores and hand grenades should be used to engage these probes. When probed, riflemen near machineguns should fire, not the machinegunners. Machinegunners must use their pistols for self-defense instead of their machineguns. When the enemy attacks, he will try to isolate friendly positions and destroy them one at a time. Well-planned, mutually supporting fires will prevent this.

COUNTERATTACK

If enemy forces penetrate a position, a counterattack is the best way to expel them. Troops in the area of the penetration must stay in their positions and continue to fire to support the counterattack. If they leave their positions while the enemy is being expelled, they increase the chance that they will be hit with friendly fires.

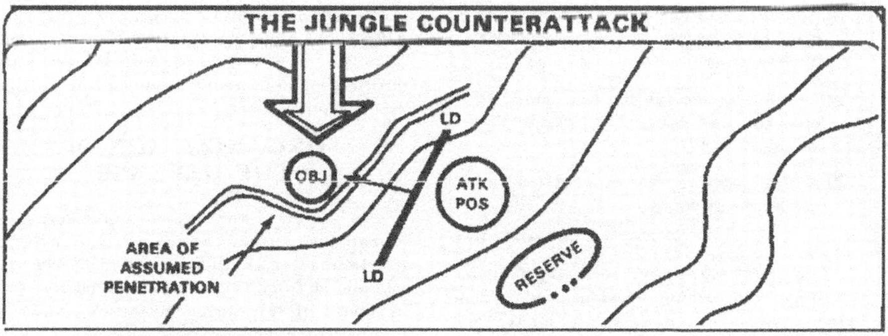

THE BATTALION COMBAT BASE

When engaged in tactical operations in the jungle, elements of the battalion will often establish a base for command and control and fire support resources, protected by a perimeter defense. These resources are called the battalion combat base.

The location of the perimeter defense to defend the battalion combat base will depend upon the:

- **Forces available to defend the combat base**
- **Ability to support subordinate units with indirect fire**
- **Defensibility of terrain**
- **Ability to communicate with subordinate units**

Prior to establishing the battalion combat base, the commander should conduct a reconnaissance to determine the defensibility of the terrain. He also plans forces required for the defense. While the defense must be capable of defeating the largest attack which the enemy is likely to conduct, it must use the minimum forces necessary. The combat support company, reinforced as necessary with an attached infantry platoon, is the largest force that is realistically available for preparing and defending the perimeter. To use a larger force would probably leave insufficient forces to fight the more important combat. Since some elements, such as the antitank platoon of the combat support company, may not be employed effectively in the jungle, they will often be available for use in the perimeter defense. The battalion commander will normally designate the combat support company or headquarters and headquarters company commander to be the battalion combat base commander and will have him take charge of the construction and execution of the perimeter defense.

Forces normally under control of the battalion combat base commander include:

- **The antitank platoon**
- **A Redeye section, if attached (both to man the perimeter and to provide antiaircraft fire)**
- **The heavy mortar platoon (both to man the perimeter and provide fire support)**
- **A rifle platoon, if provided for the perimeter defense**

The scout platoon is normally used for patrolling or screening missions, rather than being used in manning the combat base perimeter.

During construction of the perimeter defense, it is vulnerable to enemy attack. Consequently, it is imperative to complete the perimeter defense as quickly as possible and to provide maximum security during construction.

The threat of infiltration attacks must be emphasized. The enemy may not be able to conduct large scale attacks on fortified positions, but he may be capable of disrupting operations by infiltrating one- or two-man teams through the perimeter to place explosive devices on command and control facilities, artillery pieces or mortars, or ammunition storage areas. This infiltration is often preceded by a deceptive attack or probe by ground forces. Troops in the perimeter must maintain constant security, using early warning systems and continuous patrolling. Starlight scopes, OPs, unattended ground sensors, and tripflares are also used. Wire obstacles should be used to keep infiltrators out of critical facilities.

A battalion combat base may have to remain in place for a long time. Continuous firing of mortars and landing of helicopters makes concealing its location very difficult. These two factors make it necessary to harden the perimeter defense. Overhead cover and sandbagged bunkers must be provided for all fighting positions. The tactical operations center (TOO and CP should have similar protection and may also be dug underground. Mortars and artillery pieces should be dug in or fortified with sandbags.

ORGANIZATION OF THE DEFENSE

A reserve for the defense may be constituted from attachments, such as engineers (if available), or from off-shift personnel from TOC and CP elements. This reserve will react to enemy attacks, and will reinforce the defense or counterattack. They must be rehearsed on signals and actions until they become proficient. Mortars are employed to provide close-in fire support. Artillery pieces can provide direct fire but probably will not be able to provide indirect fire support of the perimeter. Hence, the perimeter should be located within range of other artillery and mortar units for additional protection.

The battalion combat base commander assigns sectors to subordinate platoons, insuring that likely avenues of approach are dominated by Dragons, machineguns, and artillery in direct fire role, if possible. He then plans indirect fires. The commander specifies points at which adjoining platoons must coordinate. The platoon leader selects each position and designates the personnel to man it.

The platoon leader must insure that he has complete coverage throughout his sector to deal with not only a mass attack but also infiltration of small elements.

Once the positions are selected, the platoon leader insures that a priority of work is adhered to.

EXAMPLE OF PRIORITIES OF WORK

Work should be accomplished in the following steps, consistent with the tactical situation and the availability of resources.

Step 1:	Air assault/ground assault seizes the site.
	Immediate security established to include OPs.
	Area swept for boobytraps.
	Mortars laid.
Step 11:	Communications established.
	CP set up.
	Position of TOC dug in.
	Selected TOC personnel displaced to perimeter defense.
Step III:	Perimeter positions established.
	Fields of fire cleared.
	Reserve force established.
	Wire to all positions.
Step IV:	Barriers and obstacles placed around perimeter defense.
	Early warning devices emplaced.
	Security/ambush patrol plans established.
	Final protective fire (FPF) fired in.
Step V:	Positions sustained.
	Positions are hardened with overhead cover.
	All other positions improved.
	More fields of fire cleared.
	Landing zone enlarged.
	Latrine, generators, and ammunition supply point (ASP) established.
	NOTE: Throughout work, camouflage must be applied.

AN EXAMPLE OF A BATTALION COMBAT BASE

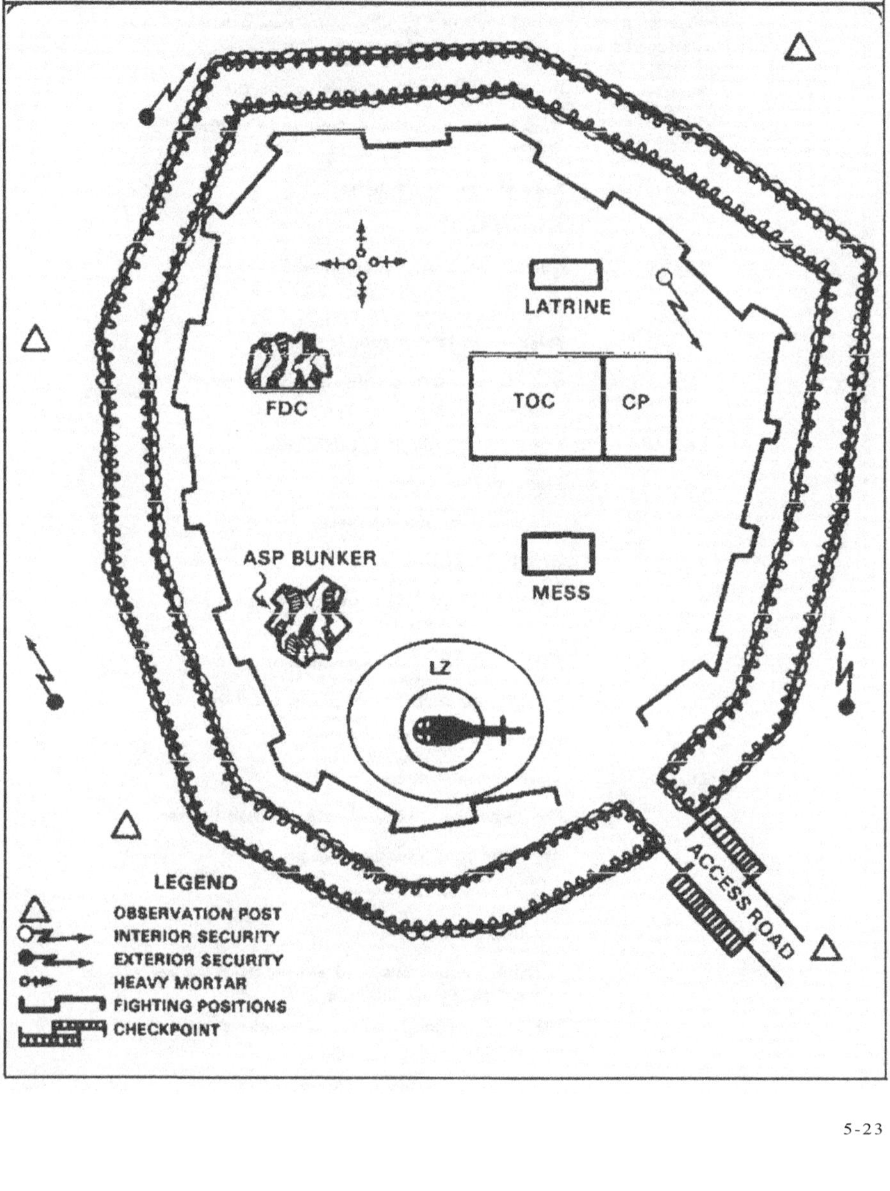

SECTION V. RETROGRADE

Retrograde operations are conducted in much the same manner as the defense, with the additional requirement to gain a mobility advantage over the enemy. This is done by taking measures to increase the mobility of the unit conducting the retrograde (reconnaissance and preparation of routes, use of helicopters, etc.), and by taking measures to decrease the mobility of the enemy (ambushes, artillery fires, mines, wire obstacles, etc.).

WITHDRAWL

Withdrawals may be conducted under enemy pressure or not under enemy pressure as explained in FM 7-8 and FM 7-10. Because of the cover and concealment provided by the jungle, a withdrawal under enemy pressure may be conducted using deception in much the same manner as a withdrawal not under enemy pressure. The ruggedness of the terrain and the strength of the attacking enemy are key factors in this type of operation. Routes, assembly areas, and new positions must be thoroughly reconnoitered. They may also be marked if such marking does not compromise security. Densely vegetated areas will require increased use of guides. Control in such areas is very difficult; therefore, leadership, planning, and rehearsal are crucial.

"J *will never again tell my platoon to withdraw • especially in the jungle - without telling it where to go. I had a hell of a time getting them together.* "

—Platoon Leader, New Guinea, World War II

Unattended ground sensors employed along likely enemy avenues of approach can be used to provide information on enemy movement and activities. This information can in turn be used to place fires on the enemy.

DELAY

The delay in sector is normally the most frequently used type of retrograde in the jungle. Ambushes may be used to halt the enemy's pursuit, and can aid the main body in breaking contact along the delay route. Mechanical ambushes, wire obstacles, and minefield will also delay the enemy. In the jungle, the delay is normally conducted in several phases: defense, withdrawal under pressure, breaking contact, and movement to and occupation of new positions where the defense starts again. As in the withdrawal, units should reconnoiter and clear routes to the rear.

Section VI. OTHER COMBAT OPERATIONS

Patrols, raids, and ambushes are nor-really used more often in the jungle than in more open terrain. The jungle permits small units to move undetected during reconnaissance patrols and achieve surprise when conducting raids and ambushes. These are small-unit operations. They depend heavily on the skill and stealth of infantry platoons and squads, and are demanding operations.

One jungle myth that was popular both during World War II and the Vietnam War was that "the night belongs to the enemy. " The enemy has no better night vision or stealth than do trained US soldiers. He will often use the night as a means to avoid US firepower, but with the proper use of patrols, raids, and ambushes, this problem can be eliminated.

"On any given night in Vietnam, American soldiers staged hundreds of ambushes, for the ambush is one of the oldest and most effective military means of hampering the enemy's nighttime exploits."

—J.A. Cash, Seven Firelights in Vietnam

Successful jungle patrols, raids, and ambushes result from detailed planning, intensive training, and constant rehearsal. Troops must be alert. A unit which has moved cross-country through the jungle until late in the afternoon will not be in a condition to succeed if it has to go out on ambush that night. Commanders must realize that such operations require time to prepare, train, and rest.

The basic techniques for patrols, raids, and ambushes can be found in *chapter 5, FM 7-8.* Specific techniques which are effective in jungle operations are described below.

PATROLS

A patrol is a detachment sent out by a larger unit to conduct a combat or reconnaissance operation. The operation itself is also called a patrol. The mission to conduct a patrol may be given to a fire team, squad, platoon, or company. The leader of the detachment conducting a patrol is referred to as the patrol leader.

CATEGORIES OF PATROLS

The planned action at the objective determines the patrol's category. There are two categories of patrols:

Combat (ambush, raid, or security) Patrol. This patrol provides security and harasses, destroys, or captures enemy troops, equipment, and installations. A combat patrol also collects and reports information, whether related to its mission or not.

Reconnaissance (area or zone) Patrol. This patrol collects information or confirms or disproves the accuracy of information previously gained.

Regardless of the category of the patrol, there are four key principles to successful patrolling. These are:

- **Detailed planning.**
- **Thorough reconnaissance.**
- **Positive control.**
- **All-round security.**

ORGANIZATION FOR A PATROL

The patrol leader decides what elements and teams are needed for his patrol, selects men or units for these elements and teams, and decides what weapons and equipment are needed. He should, however, use his unit's normal organization (squads and platoons) and chain of command (squad and platoon leaders) as much as possible to meet these needs. For example, a combat patrol may be organized like this: the company headquarters is the patrol headquarters; the 1st platoon is the assault element; the 2d platoon is the security element; and the 3d platoon and weapons platoon make up the support element.

GENERAL ORGANIZATION

A patrol generally consists of a patrol headquarters and the elements needed for the mission.

Patrol Headquarters. The headquarters (HQ) of a company-size patrol normally consists of the same number of men as a regular company headquarters. However, regardless of a patrol's size, its leader tailors the headquarters to meet mission needs. The patrol headquarters has the same responsibilities as any other command element.

Reconnaissance Patrol. In an area reconnaissance (recon), a patrol has a reconnaissance element and a security element. In a zone reconnaissance, a patrol has several reconnaissance elements. Each one provides its own security.

Combat Patrol. A combat patrol normally has an assault element, a security element, and a support element. At times, the support element may be omitted by combining it with the assault element.

In general, jungle terrain affords excellent concealment, provides some cover, and hinders enemy observation and movement. During rainy periods, the sound of movement is less obvious. Wet ground and wet vegetation also muffle noise. All of these factors favor the patrolling unit.

On the other hand, the difficulties of movement and control and the ease with which the enemy can infiltrate friendly units are disadvantages to units patrolling in the jungle. These factors can best be overcome by training and discipline.

Silence, in both voice and movement, is essential at all times during a jungle patrol. With practice, it is possible to move steadily, deliberately, and carefully through the jungle, parting the undergrowth instead of crashing through it or cutting through it with machetes. Troops should avoid walking on dry leaves, sticks, rotten wood, or anything that would make noise. Machetes should be used to cut trails only as a last resort. Talking should be done in a whisper, and arm-and-hand signals should be used whenever possible.

Trails should be avoided. Patrols should make every effort to hide signs of movement, especially when moving through untraveled territory or near enemy positions.

Some techniques which may be used include:

- **Requiring all troops to wear boots that have the same pattern on their soles.**
- **Requiring troops to carry only the mission essentials, and do not let them litter.**
- **Cautioning troops to avoid small saplings, when going up hill. The shaking of overhead branches can be seen and heard at a distance.**
- **Requiring troops to keep off trails. If necessary to monitor or guide on a trail, patrols should move parallel to the trail and not on it.**

Native scouts are valuable in patrolling because they are often very familiar with the terrain. Patrol leaders must realize, however, that a scout's function is only to show direction and provide information. He should never lead the patrol. The correct position of a scout is with the patrol leader, so that the leader can make decisions based on the scout's advice.

RAID

Raids in the jungle environment must be keyed to reliable intelligence. The actions of the raiding unit must be decisive and rapid in order to catch an elusive jungle enemy. A raid's success depends on good intelligence and a sound plan.

Jungles favor raid operations. The excellent concealment enables skilled raiding patrols to operate deep in enemy territory. Platoon-sized units are best suited to jungle raids. Supporting artillery fires should be planned, but due to difficulties of control, timing, and communications, jungle raids may be executed without artillery support. Surprise is a key ingredient of a successful raid.

Raids that require deep penetration into enemy-held areas are best executed by establishing a patrol base in the general area of the final objective. From there, reconnaissance patrols can be sent to scout enemy positions while the remainder of the force completes its preparations for the raid. Helicopters and watercraft are effective means of transporting a raiding force rapidly to the vicinity of its objective without depleting their physical strength in a difficult march.

AMBUSHES

The ambush is more important, more effective, and more frequently used in jungle fighting than in any other type of combat. Jungle terrain provides many opportunities for a well-concealed force to gain surprise. Surprise is essential for a successful ambush.

Destruction of enemy forces is the primary purpose of most ambushes, but other benefits result from a well-executed ambush program.

These benefits include:

- **Disruption of enemy operations, since troops become reluctant to move and fight in areas where ambushes are frequent.**
- **Capture of prisoners and equipment which may yield intelligence data.**
- **Capture of supplies, thus increasing combat effectiveness at the expense of the enemy. In some instances, this is the primary source of supplies for guerrilla forces.**

More than in any other type of terrain, jungle ambushes require high standards of discipline. Soldiers on an ambush must be prepared to remain in the same position for hours at a time, without being able to sleep, talk, or smoke. They must endure insects and resist the desire to make any quick moves to swat or brush the insects away. All these require extensive training to develop the patience and self-discipline required.

The location for an ambush should be chosen after a careful analysis of the terrain, using maps, aerial photographs, and personal reconnaissance. The site chosen must contribute to the surprise of the ambush. Many times the selection of a site for surprise alone will be more effective than attempting to ambush from a site which is in other respects tactically sound but at which the enemy is sure to be suspicious. Covered routes of approach and withdrawal, good fields of fire, and canalization of the enemy are characteristics of a good site. The site should always be reconnoitered and approached from the rear.

In no other operation is camouflage more important than in the ambush. Weapons should fire through screens of undisturbed, living foliage. Spoilage resulting from the preparation of positions must be removed from sight. There can be no unnecessary noise or movement. If reliefs are used, they should be scheduled so that only a few men move at any one time. One or two men moving are harder to detect than an entire relief moving at once.

DEFENSE AGAINST AMBUSH

Since ambushes are more frequent and effective in the jungle than in any other type of terrain, a unit moving through the jungle must take all possible measures to reduce its vulnerability to ambush. The most effective means of countering an ambush is to detect it before entering the kill zone. This, however, is not always possible.

Dismounted troops have an advantage over mounted troops in avoiding ambushes because they do not have to move on roads or trails. Commanders of dismounted units should make a map and aerial photograph reconnaissance to detect likely ambush sites and plan routes which avoid them. During movement, security to the front, rear, and flanks should be maintained at all times. Alert troops, good noise discipline, and well-rehearsed signals are other means which will reduce the chances of ambush. Accurate land navigation, continuous fire support planning, and counterambush drills are also important antiambush techniques.

Mounted troops are very vulnerable to jungle ambushes, especially where the foliage grows up to the edge of a road. Ambush of vehicular columns traditionally has been a primary tactic of jungle enemies. As a result, traffic in jungle areas must be tightly controlled and kept to a minimum. All vehicles should have armed riders. Armored vehicles should escort convoys, and traffic information should be carefully guarded. Fire planning and route selection and reconnaissance are important for mounted troops as well as for dismounted troops *(app M, FM 7-8).*

PREPARATION OF TRUCKS FOR MOVEMENT

To prepare trucks for jungle convoys:

1. Lower windshields and cover them with tarps (this reduces danger of flying glass and stops glare).
2. Place tailgates halfdown to ease dismounting (troops can then shoot and dismount fast).
3. Put at least two layers of sandbags on the floors of the cabs and cargo areas to protect against mines.
4. Stack sandbags along the sides of the cargo beds to protect troops from small-arms fire from the flanks.
5. Stack and fasten sandbags around the engine compartments to protect the engines (do not block the flow of air to radiators).
6. Place troop seats in the center, facing out, so troops can react fast.
7. Install wire-cutting apparatus in front of trucks to prevent injury of troops from wire stretched across the road.
8. Remove the rear safety straps for fast dismounting.

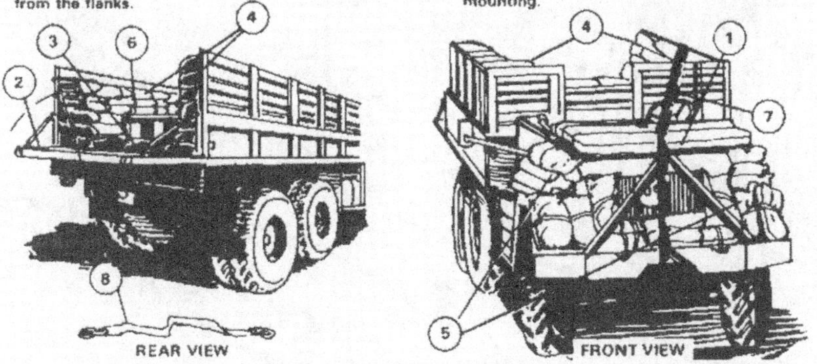

COUNTERAMBUSH MEASURES

Troops must also be trained in counter-ambush measures. The key is early detection followed by reflex-type counteraction, a high volume of return fire, and relentless pursuit. The most effective counterambush measures are well-rehearsed immediate action drills. Every soldier must know exactly what he is supposed to do.

Dismounted troops should react to an ambush immediately, firing into the ambushers without orders. Building and retaining fire superiority is the best initial defense against an ambush. If a patrol finds itself in an enemy ambush, it must get out of the kill zone immediately. It must take the following immediate actions:

Troops in the kill zone, without order or signal, immediately return fire, and quickly move out of the kill zone by the safest way. (There is no set way to do this; it must be each soldier's decision for his situation.) Smoke grenades can help conceal the troops in the kill zone.

Troops not in the kill zone fire to support the withdrawal of the troops in the kill zone.

If a dismounted patrol is ambushed, it should attempt to break contact and reorganize in the last designated rally point.

Mounted troops who are ambushed should attempt to drive rapidly out of the kill zone. Vehicles approaching the kill zone should stop so they do not enter it. Troops should then dismount and maneuver to destroy the ambush.

In any case, the rapid call for supporting artillery and mortar fire will help the ambushed force to gain fire superiority and will assist the maneuver to destroy the ambush.

REACTING TO AMBUSH

CHAPTER 6

Helicopter, Armor, Mechanized Infantry, And Combat Support Operations

Section I. GENERAL

In the jungle, infantry battalions are the heart of both the offense and the defense. These battalions are normally supported by a mixture of helicopter, armor, artillery, air defense, engineer, and other units.

This chapter outlines those techniques which are most effective in supporting infantry engaged in jungle operations.

Section II. Helicopters

HELICOPTER ROLES

The helicopter is a combat system that is unaffected by the poor trafficability of jungle areas. As a result, it has become a vital part of US jungle operations. US forces have used helicopters in the jungle successfully for heliborne fire, reconnaissance, air assault, command and control, resupply, and medical evacuation. This section will discuss heliborne fire, reconnaissance, air assault, and command control in the jungle. *For resupply and medical evacuation, see chapter 7.*

In all of these operations, bad weather and enemy air defense are major considerations in the jungle as they would be anywhere else.

CONTENTS

Heliborne fire is provided by the attack helicopter unit.

> **The attack helicopter unit performs three basic missions:**
>
> - **Overwatch and security**
> - **Engagement of targets located by ground elements**
> - **Independent target engagements**

Attack helicopters in an overwatch and security role protect other assets, such as other helicopters or convoys. Attack helicopters in the heliborne fire role can augment the fires of units fighting on the ground. In jungle operations, they maybe the only additional means of fire available. Finally, attack helicopters may acquire and engage targets independent of any other operations.

HELICOPTER ARMAMENT

Although the Cobra, AH IS, armed with the TOW, has become the standard US attack helicopter, the nature of the typical jungle enemy and the terrain make the 2.75-inch rocket (HE) a more useful weapon against known point positions. The "Flechette" 2.75-inch rocket, 7.62-mm minigun and 40-mm grenade launcher, also on the attack helicopter, are extremely effective against dismounted troops.

COMMUNICATIONS

The most important consideration when using attack helicopters to augment the fire of ground troops is communications. Both radio and visual communications means must be established and maintained. Radio communications are used to pass target information from the ground commander to the helicopter. Visual communications are used to mark the location of friendly troops. Common daytime signals are a smoke grenade, a panel, and a mirror. A flashlight, a strobe light, and a chemical light stick are effective for signaling at night.

COMMUNICATION WITH ATTACK HELICOPTERS

While the helicopters are en route to the target area, a member of the attack helicopter team will contact the ground commander over the radio. At this time, the ground commander must transmit *target handoff information.*

TYPICAL EXCHANGE OF TARGET HANDOFF INFORMATION

IDENTIFICATION	Helicopter: —	"55T7 this is 24J6."
	Ground Commander: —	"24J6 this is 5517."
ENEMY INFORMATION	Ground Commander	"I am pinned down by an estimated enemy company. I am receiving AK-47, RPD, and RPG fires."
OWN SITUATION	Ground Commander:	"I am in a perimeter defense. There is no artillery impacting at this time. I will mark my location with smoke."
	Helicopter	"I'm about 30 seconds from your reported location. Pop a smoke grenade."
VISUAL IDENTIFICATION	Ground Commander: —	"WILCO."
	Helicopter: —	"I identify yellow smoke."
	Ground commander	"Roger, yellow smoke. Enemy is at azimuth 120 degrees from yellow smoke, estimated distance 200 meters."
COORDINATION	Helicopter:	"Roger, I'll make my run from northeast to southwest."

RECONNAISSANCE

Reconnaissance missions are normally given to air cavalry. Air cavalry troops and squadrons are equipped with attack helicopters, scout helicopters, and utility helicopters. The air cavalry troop is organized with a troop headquarters, aeroscout platoon, aeroweapons platoon, aeroreconnaissance platoon, and service platoon. The troop is task organized for specific reconnaissance missions.

The aeroscouts use terrain flying while looking for signs of the enemy. Attack helicopters from the aeroweapons platoon support them and engage targets as they are acquired. If a thickly vegetated area cannot be reconnoitered from the air, the aeroreconnaissance platoon can be inserted on the ground by the utility helicopters.

AIRMOBILE OPERATIONS

This employment of air cavalry is effective in the jungle. The information gathered can be rapidly processed into intelligence. This in turn can be used to plan timely commitment of troops into the area.

Airmobile operations in the jungle are the most rapid means of concentrating combat power at a critical time and place on the ground. The doctrine for airmobile operations is found in FM 90-4, and it is generally applicable to jungle operations.

Special considerations in jungle operations include:

- **Utility helicopters will not be able to lift the same size loads that they can in more temperate areas. Sometimes this may result in as few as 5 soldiers per sortie for UH-1H and 19 soldiers for UH-60.**
- **Radio and visual communications between ground and air are vital to insure proper coordination. The thick jungle foliage will often make this more difficult than in other types of terrain.**
- **There maybe few suitable landing zones (LZ). Most landing zones will be only large enough to support one or two helicopters at a time.**
- **Units may have to be resupplied totally by air.**

One of the most common uses of the helicopter in jungle operations is command and control. Command and control (C&C) helicopters are specially outfitted with radios which allow communications with elements on the ground, supporting artillery, and other aircraft. These helicopters also have seats in them arranged so that commanders, S3s, and fire support coordinators (FSCOORD) can talk to each other and perform their respective functions. Such helicopters are normally available to command groups down to and including infantry battalion level.

To use the C&C helicopter effectively, a commander should exploit its unique capabilities:

- **The helicopter's speed enables a commander to influence the action in widely separated areas.**
- **The helicopter's powerful radios and flight altitude permit the commander to communicate more effectively and reliably than he could on the ground.**
- **The unrestricted observation from the air allows the commander to see the terrain better, adjust fires more accurately, and control attack helicopters more effectively than he could on the ground.**
- **When used for a commander's reconnaissance, the C&C helicopter is a significant help in planning operations. Likewise, a battalion commander who takes his company commanders in the C&C helicopter to issue an operation order (OPORD) can use the terrain below as a reference— like a readymade sandtable.**

The C&C helicopter can be one of the jungle commander's greatest assets. It can also be abused. The keys to its effective use are a few common sense considerations.

The helicopter cannot become a substitute for command presence on the ground. (A commander cannot merely fly over a unit and be satisfied that he knows its situation. The view from the air often gives a distorted picture of conditions on the ground.)

The helicopter should not interfere with ground operations. (If a unit has to devote time to cutting an LZ every day for a visit from the commander, it will get little else done.)

The helicopter must not violate a ground unit's security. (A helicopter hovering overhead soon reveals the location of a unit on the ground.)

The helicopter must not intrude on a ground unit's internal radio nets. (The helicopter's radios are more powerful, and will interfere with the ground commander's ability to control his own unit.)

The helicopter is vulnerable to ground fire. (Although the nature of the air defense threat will be different in every conflict, no commander can expect to survive if he maintains a stationary position anytime in the vicinity of ground contact or orbits at high altitudes. The helicopter is extremely vulnerable to small-arms fire and heat-seeking missiles.)

Section III. ARMOR OPERATIONS IN THE JUNGLE

Up to this point, this chapter has covered infantry tactics in the jungle. Armor may also play a role on the jungle battlefield. Because the tank's mobility is greatly restricted on jungle terrain, its role is different in the jungle than that on more open terrain. Instead of rapid envelopments and deep penetrations, tanks primarily provide fire support for infantry in jungle fighting.

Tanks can perform the following missions in the jungle:

- **Support attacking infantry. In the meeting engagement, they move to the point of contact to provide direct fire support.**
- **Provide heavy fire support in the defense and conduct counterattacks.**
- **Support infantry in retrograde operations.**
- **Secure convoys and protect lines of communications and key facilities.**
- **Attack or defend against enemy armor.**

TACTICAL OPERATIONS

OFFENSIVE OPERATIONS

In the movement to contact, infantry will normally lead, supported by tanks from overwatch positions whenever the terrain permits. Terrain and vegetation will dictate distances between the forces, but the tanks should be close enough to the infantry to be able to move forward quickly to provide immediate fire support.

In a meeting engagement or attack, once the infantry makes contact, tanks move to positions from which they can support the infantry.

TANKS IN POSITION TO SUPPORT INFANTRY

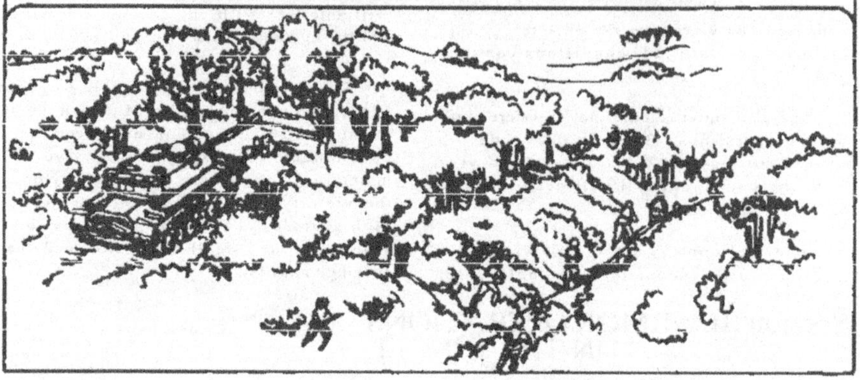

Tanks can fire their machineguns and main guns using high explosive antitank (HEAT) and antipersonnel rounds to destroy or suppress the enemy, clear jungle foliage, and destroy fortified positions.

Should the enemy have armor, the tanks can more effectively attack the enemy armor than can the TOW or Dragon antitank guided missile because the tank gun can better fire through the jungle foliage.

When terrain permits, infantry can fix the enemy, while tanks create a penetration to split the enemy defense. The tanks and infantry can then destroy the severed forces.

DEFENSIVE OPERATIONS

Tanks are used in the jungle defense as they are used in any other environment. They add greatly to close defensive firepower and serve as a mobile counterattack force.

In the position defense, tanks should be positioned to:

- **Block possible armor or motorized penetrations**
- **Move quickly to fill gaps in the defense**
- **Provide antiarmor fire**

Tanks in the jungle defense are vulnerable to infiltrators; consequently, all-round security and close coordination with infantry are essential.

Hurricane fencing should be erected as a screen in the defensive position to provide stand-off against enemy use of rocket-propelled grenades (RPG) and antitank guided missiles (ATGM).

RETROGRADE

Tanks can support retrograde operations in much the same way as in the defense. Contrary to general practice, in jungle retrograde operations tanks will often disengage first, covered by the infantry.

LIMITATIONS

RESTRICTIONS ON USE OF ARMOR

Vegetation. Densely forested areas, where tree trunks are close and heavy above-ground root systems exist, may make movement difficult. Thick stands of bamboo may slow or even stop tanks.

Topography. Many jungles exist in rugged mountainous areas which will impede tanks. Swamps, coastal river basins, and other areas intersected by waterways pose obstacles to tanks because of their soft soil and frequent deep channels.

Weather. Rainy seasons (monsoons) cause rivers and streams to rise and become unfoldable. Heavy rains may also cause damage to roads.

TECHNIQUES TO OVERCOME ARMOR RESTRICTIONS

All of these factors limit but do not exclude the use of tanks.

A commander can overcome those limitations by using these techniques:

- **Know the terrain. Know where tanks can travel. Avoid areas which are obviously impassable.**
- **Know the weather. What were recent conditions? What is expected? What effect will these conditions have on the use of tanks?**
- **Move dismounted infantry in front of tanks on unfamiliar and heavily vegetated terrain. They can check and verify conditions, act as guides, and provide security.**

TIPS FOR MOVEMENT IN THE JUNGLE

Jungle conditions vary greatly from place to place and season to season.

Here are some movement tips which apply to most jungle areas:

- **Red silt soils tend to break down quickly when wet. They may support a single tracked vehicle but may become untrafficable with heavy use.**
- **Inundated areas containing yellowish reeds and cloudy water usually have bottoms too soft to support tanks.**
- **Rice fields, in the dry season, are usually trafficable to tanks and personnel carriers. During the rainy season, they may be untrafficable. Fields with standing water in the wet season may have a bottom too soft to move on; those containing clear water and green vegetation usually are on firm ground and are trafficable. One technique for determining if rice fields are trafficable is "poling." Using this technique, troops precede tanks across the field they are to cross, sinking poles (1 to 1 1/2 inches in diameter, with flat ends) into the ground. If the poles cannot be sunk, the field is usually firm enough to cross.**
- **River and stream bottoms usually are untrafficable. The armored vehicle launched bridge (AVLB) can span 17 meters (57 feet) and is more than adequate for most stream crossing. Care must be taken to insure that the shoulders of the banks can support the AVLB while tanks cross. When the AVLB is not avilable, perforated steel planks can be used to provide a firm surface on which tanks can ford small streams.**

Section IV. MECHANIZED INFANTRY

Mechanized infantry units are able to provide a commander fighting in a jungle environment a greater flexibility than that afforded by nonmechanized infantry. Mechanized infantry may be employed as *a mounted force, a dismounted force, or airmobile infantry.*

CAPABILITIES

OFFENSIVE

In offensive operations, mechanized infantry may conduct the following offensive actions:

- **Movements to contact**
- **Attacks-hasty or deliberate**
- **Reconnaissance-in-force**

DEFENSIVE

In defensive operations, mechanized infantry may be used:

- **In a strongpoint defense**
- **As part of a linear defense**
- **To establish a perimeter defense**

OTHER OPERATIONS

Mechanized infantry units, especially when formed into tank-infantry company teams, may also be used in other operations such as:

- **Route clearance and security**
- **Convoy escort**
- **Reserve of a higher headquarters (either for offensive or defensive operations)**
- **Raids**

CONSIDERATIONS FOR EMPLOYMENT

As mechanized infantry may be used in different ways, the precise mission it is assigned must be weighed against its strengths and weaknesses in a jungle environment.

STRENGTHS	WEAKNESSES
■ **Quick reaction over distances.** The unit has its own transportation immediately available, and considerable firepower to accompany it. In savanna or on firm but lightly vegetated ground, mechanized infantry is able to move considerable distances in a relatively short period of time.	■ **Trafficability.** When used as a mounted force, mechanized infantry units may be limited in the scope of their operations by the terrain in which they are employed.
■ **Firepower.** A mechanized infantry force has greater firepower than other Infantry units. The carrier's caliber .50 machinegun can be used to strengthen defenses or provide overmatching fires.	■ **Noise.** The operating noise of the tracked vehicles limits the ability of mechanized infantry units to achieve surprise while moving.
■ **Large load capacity.** The armored personnel carrier (A PC) provides mechanized infantry units larger capacity to carry food, water, and ammunition than other units.	■ **Logistics requirements.** Mechanized infantry units increase logistics requirements, because of increased petroleum, oils and lubricants (POL) consumption; greater repair parts requirements; more frequent recovery operations; and larger maintenance requirements.
■ **Communications.** A mechanized infantry unit has a communications advantage because of both the number and types of radios it uses.	

Disposition of MI 13 APC when troops are dismounted. When employing mechanized infantry as an airmobile force or as dismounted infantry, consideration must be given to the disposition of the M113 APCs. It is usually advisable for a driver to remain with each vehicle. In addition, some infantrymen may have to provide security for the carriers and thus will not be able to fight with the dismounted element.

MECHANIZED OFFENSIVE OPERATIONS IN A JUNGLE ENVIROMENT

The offensive operations that mechanized infantry units are most likely to conduct in a jungle environment are movement to contact and reconnaissance-in-force.

A movement to contact is used to gain or regain contact with the enemy, and to develop a situation. It serves as the first stage of operations against an enemy force which has not been located, and normally ends in a meeting engagement.

A reconnaissance-in-force is employed to force the enemy to react so that friendly elements can develop information about the enemy by discovering his location, disposition, and intent. It is used when other means of gaining information about the enemy are not available.

Hasty and deliberate attacks are conducted in much the same manner as they are in conventional terrain. The techniques, however, may differ due to the jungle environment, especially the strict application of control measures in controlling maneuver elements in the thick jungle vegetation. Also, an attack will normally be conducted on a more narrow front due to limited visibility.

MOUNTED MOVEMENT

Hasty and deliberate attacks are conducted in much the same manner as they are in conventional terrain. The techniques, however, may differ due to the jungle environment.

When moving in the jungle, either cross-country or along roads, it is best to move in multiple columns on as broad a front as possible. The lead elements should be preceded by an aerial route reconnaissance. Indirect fire support must be closely coordinated and instantly available, as contact is often violent and without warning. TACAIR and attack helicopters provide another responsive means to react to enemy contact.

DISMOUNTED MOVEMENT (APC ACCOMPANYING)

When in contact or contact is anticipated the troops dismount and fight on foot. The caliber .50 machinegun on the APC can provide fire support maneuvering elements.

RECONNAISSANCE BY FIRE

Movement in the jungle carries with it the prospect of ambush at any time. Reconnaissance by fire is a technique that can be used to decrease this danger. In this technique, lead elements engage suspected or likely enemy locations with automatic weapons as the unit moves. The enemy is then made to move or return fire. When considering a reconnaissance by fire, the commander should realize that it uses a lot of ammunition and reveals his location to the enemy. Reconnaissance by fire is not an excuse for indiscriminate fire by individuals.

MOVEMENT TO CONTACT

Mounted movement to contact in the jungle is characterized by deliberate use of traveling overwatch and bounding overwatch. The overwatch positions will be closer than in other areas, due to the vegetation and terrain. Contact with the enemy will usually be made at close range (within 200 meters); therefore, automatic weapons should be kept well forward. Frontal, flank, and rear security may be established mounted, but more often should be provided by dismounted elements. The security elements should be rotated frequently to prevent fatigue.

RECONNAISSANCE-IN-FORCE

When conducting a reconnaissance-in-force, a mechanized infantry unit may have lead elements conduct a mounted movement to contact. At likely ambush sites along the route, however, these elements must dismount and patrol. These patrols are normally conducted by a platoon's maneuver element. The earner teams, usually with two men in each, should overwatch the maneuver element. They can be called forward when required.

DELIBERATE ATTACKS

Deliberate attacks in the jungle are usually conducted dismounted, with Ml 13s supporting the attacking elements. Care must be exercised to insure that the maneuvering infantry does not mask the supporting fires during the assault. The assault can be preceded by artillery preparation and use of close air support.

MECHANIZED INFANTRY DEFENSIVE OPERATIONS IN A JUNGLE ENVIROMENT

Mechanized infantry units are a formidable force when defending. The most common types of defensive operations they might participate in are forming a defensive perimeter, establishing a strongpoint, or taking part in a position defense.

DEFENSIVE PERIMETERS

Defensive perimeters may be established at any time but are normally used during periods of limited visibility to increase the security of the force and to allow time for maintenance and rest. The perimeter is a hasty defense technique. It is usually only a temporary arrangement and is moved frequently. A perimeter is not oriented against a particular enemy force, but takes advantage of terrain to obtain the greatest security possible.

Establishment of a Perimeter. An M113 leaves a track that is difficult to hide and easy to follow. Therefore, a mechanized infantry perimeter is difficult to conceal. Consideration should be given to the placement of ambushes along the back track path leading into the perimeter. The perimeter is very similar to that used by dismounted infantry but it is adjusted to accommodate the tracked vehicles. It is best to have a quartering party precede the unit but this may not be possible in all cases. A unit SOP should be the basis for specifying the establishment of a defensive perimeter. M113s may be used to clear fields of fire where they are capable of doing so. (An APC is capable of clearing considerable brush in a short period ot time.) The M113s are then parked in a circle, wagon train style, facing out. Individual fighting positions are prepared in front of the parked vehicles. Dismounted observation posts (OP) and ambush patrols are sent out. These elements may use Claymore mines and early warning devices. The caliber .50 machineguns may be dismounted or remain on the Ml 13s but are manned at all times. When the soldiers are allowed to sleep, they should be in or close to fighting positions in order to be near overhead protection. Ramps should normally be raised, with access to the M113 through the troop door, in order to provide protection to the crew from indirect fire.

Defense of a Perimeter. An enemy ground attack against a perimeter defense maybe by a surprise assault or be preceded by preparatory fires. The unit should continually have all the automatic weapons manned and ready to fire, *but care must be taken to insure patrols and OPs are back within the perimeter before machinegun FPLs are fired.* M113s should not be moved during a night attack on a defensive position, except under emergency conditions, due to the danger of injuring prone friendly troops.

STRONGPOINT DEFENSE

A strongpoint is a defensive position which is fortified as extensively as time and materials permit. It is normally located on a terrain feature critical to the defense. Individual fighting positions should be prepared with overhead cover. The M113s should be placed in hull defilade positions, which are formed by natural terrain features or dug by a bulldozer. Each position should be deep enough to protect the vehicle. The caliber .50 machinegun may be dismounted or remain on the vehicle. If it remains on the vehicle, the position should allow the caliber .50 machinegun to attain grazing fire. If materials are on hand and time is available, a chain link fence may be emplaced 10 to 15 meters in front of each Ml13 position to cause premature detonation of antitank rounds. Each squad's position and OP should be linked with the platoon leader in a telephone hot-loop, with radio used as an alternate means of communication.

SUGGESTED DEFENSIVE TECHNIQUES FOR MECHANIZED INFANTRY UNITS

Each Ml13 can carry a large amount of equipment. The mechanized unit should take advantage of this by habitually carrying equipment to aid in preparing defensive positions.

This equipment may include:

- **Concertina and barbed wire**
- **Engineer stakes**
- **Pierced steel planking**
- **Sandbags**
- **Chain saws**
- **Pioneer tools**
- **Hurricane fencing (RPG/ATGM screen)**
- **Night vision devices**

M60 machineguns should be dismounted wherever possible and placed in prepared positions. The caliber .50 machinegun may be used either mounted or dismounted. It should be dismounted when the terrain prevents the gunner from getting a suitable mounted firing position. During mounted movement, the caliber .50 machinegun should be manned by the track commander and the M60 machinegun manned in the cargo hatch. *(See FM 7-7.)*

OTHER MECHANIZED OPERATIONS

When operating in the jungle, mechanized infantry units may be given missions which they would not normally perform in other types of terrain. Such missions require special planning and coordination before and during the conduct of the mission.

ROUTE CLEARANCE AND SECURITY

These operations are conducted when a route has been closed or unused. They are usually conducted so that supply activities may occur, but should not be confused with convoy escort operations, which require different tactics and techniques. Route security and clearance operations are oriented on a specific route and the surrounding areas, to insure that vehicular operations are not interrupted along that route.

Route clearance operations are conducted to eliminate the enemy along the road and to remove any explosives which may have been placed there. Whenever possible, route clearance is a combined arms effort involving as a minimum the use of armor, infantry, engineers, artillery, and Army aviation. Route clearance involves deliberate, detailed, and coordinated actions which are slow. The route must often be walked by mine-sweep teams, and the areas adjoining the route must be cleared by dismounted infantry. During route clearance operations, the mechanized infantry elements normally provide security forces for the combat engineer mine-sweep teams. The infantry is also used to clear the areas on either side of the road.

Route security missions are characterized by continuous activity to prevent the enemy from cutting the route or ambushing elements using it.

Patrolling is the key to route security, day and night. Patrols should be dispatched so there is no pattern. Patrols are usually squad-sized and are assigned specific areas of responsibility. They may at times move mounted but dismount where appropriate. These areas of responsibility should extend to 1 kilometer each side of a route, for the mission is to prevent enemy mining or ambushes. A reserve force should be centrally located to be able to react in any direction.

Bridges or large culverts along roads are prime locations for enemy demolitions and ambushes. They must be kept clear and intact. A good way to do this is to ambush near them at night, and patrol from them during the day. Night observation devices should be used in ambush operations.

CONVOY MOVEMENTS

Coordination must take place before and during a convoy movement. Each patrol must know the time of entry and time of exit of each convoy, the numbers and types of vehicles, and whether the convoy has an armed escort. Just prior to the convoy's approach, the security patrols intensify patrolling the areas immediately next to the road, and assume security positions along the known or suspected enemy avenues of approach. The convoy commander should have each patrol leader's callsign and frequency, and notify him when he is entering and leaving the area.

The elimination of the enemy is the responsibility of the unit through which the convoy is passing. If the convoy makes contact with the enemy, it should get out of the contact area as quickly as possible. The convoy commander will call the appropriate patrol leader or area commander and inform him of the contact. It is the area commander's responsibility to relieve the convoy, reinforce friendly elements, and coordinate indirect fire or air support. The convoy, or its escort, fires on ambush locations until they can get out of the kill zone.

Convoy Escort. The purpose of a mechanized infantry convoy escort is to move with and secure a convoy. The convoy commander establishes the route, checkpoints, fire coordination, and communications. He supervises rehearsal of actions on contact. He also identifies critical loads in the convoy and priorities for cross-loading. The convoy commander and escort should operate on a common radio frequency. Cross-loading spreads critical cargo in more than one vehicle. Convoys can use aerial reconnaissance aircraft as a part of the escort.

The mechanized infantry in their M113s and tanks, if available, both lead and follow the convoy. Wherever possible, the M113s will travel in pairs. If tanks are present, they normally lead the M113s. Communications among all convoy elements are essential. The convoy will travel according to the rate of the slowest vehicle.

The suppression and elimination of hostile fire is the responsibility of the escort commander. His first priority is to preserve the convoy and its contents. Upon being engaged by the enemy, the convoy will speed up to get out of the kill zone. Accompanying mechanized infantry and armor will assault by fire first, and maneuver as necessary. Any activity other than in the immediate area must be coordinated with the commander in whose area the ambush occurs, particularly if indirect fires or pursuit becomes necessary.

TANK-INFANTRY TEAMS

Tanks and mechanized infantry are often restricted by jungle terrain, but where they can be employed, they complement each other well. For this reason, mechanized infantry and armored units are often teamed. The two most common uses of tank-infantry teams in the jungle are for movements to contact and for convoy escorts.

Each must know how the other operates and the characteristics of all weapons systems. The tankers must always know where dismounted infantrymen are and the infantry must protect the tanks from close-in enemy attack.

VEHICLE RECOVERY

Mechanized infantry's primary advantage is its ability to move rapidly as compared to nonmechanized infantry. This often will be curtailed by jungle terrain. Every effort must be made to prevent M113s from becoming stuck or sustaining mechanical malfunctions that render them inoperative. The two most common occurrences are throwing a track and getting stuck in mud. The easiest way to prevent either is to precede each M 113 with a ground guide.

If an M113 becomes stuck, the primary recovery vehicle is the M578 vehicle track retriever (VTR). One is assigned to each company. It is equipped with a crane and boom, chains, and various other equipment, and it is usually located in the company trains. *It should be escorted to and from a recovery site as it is not a combat vehicle.* Every effort should be made to recover vehicles by using another M113 before the M578 is used, unless an M578 is immediately available. The M578 should be moved only when needed because it has less trafficability than the M113, weighs more, and is wider, which make movement in the jungle difficult for the M578.

Thrown Tracks. An M113 usually throws a track when the track is broken or detached from the drive sprockets. Repair time depends on a number of variables but usually varies from 45 minutes to 4 hours. The best preventive for thrown tracks is maintenance and good driving techniques. These involve keeping the track in good repair and under proper tension, the drive sprockets turned or changed, and supervising the drivers.

Mired M113s. Despite the M113s ability to move over most terrain, getting stuck in mud is not uncommon, especially when operating near rice fields or in areas of high rainfall. In these circumstances, ground guides are of little help. Infantry squads must develop expertise in expedient recovery methods. Each M113 should carry tow cables and also chains. Since M578s will not always be available, the squad must know how to recover the vehicle without VTR support. In such instances, other M113s pulling in tandem may be necessary. A tank is even better because it is much heavier than an M113, is more powerful, and generally has greater traction, except in mud.

LOAD PLANS FOR THE M113

Load plans are designed to standardize loading of equipment and supplies and evacuation of priority items if the M113 is rendered inoperable. The plans further serve to standardize operations within a unit. Load planning requires a reverse sequence—that is, items that are used infrequently and are less critical to the mission than others are loaded first. Items that are used on a daily basis and are essential to mission accomplishment must be readily available. When planning loads, consideration must also be given to maintenance. For example, radios should be easy to get to, and the engine compartment should not be blocked. The load plan diagram should also establish a numerical or alphabetical priority of equipment and supplies.

Section V. COMBAT SUPPORT

RIFLE COMPANY MORTAR SECTION

The rifle company's mortar section's mission is to provide close and continuous indirect fire support to the company. Where trafficable road networks are available, the mortar section can displace to perform its missions while mounted. It may, however, have problems in the jungle environment due to a limited number of suitable firing positions, as a result of heavy rains and restricted overhead clearance. The use of chain saws and demolitions may be necessary to obtain mask and overhead clearance. It also may be possible to fire from positions on or near roads and trails.

When the company operates in terrain where vehicles cannot go, several difficulties surface. Moving a dismounted mortar section in rugged jungle terrain with all of its weapons components and ammunition is difficult. The infantry platoon can be tasked to assist; however, this will severely hamper their ability to maneuver. If the mortar fire is absolutely essential to the company's scheme of maneuver, then consideration should be given to the employment of only one mortar. Also, consideration should be given to the displacement of mortars by helicopter.

Ideally, the rifle platoons will operate within range of the mortars. Resupply is provided by vehicle or helicopter. If the company is operating within range of a 4 .2-inch mortar platoon or an artillery battery, then consideration may be given to using the mortar section as infantry. This will allow the company more operating flexibility and considerably lighten the soldier's fighting load.

FIELD ARTILLERY

The mission of the field artillery is to provide close and continuous indirect fire support; destroy, neutralize, or suppress the enemy indirect fire; and coordinate all supporting fires. A jungle environment poses unique challenges to both maneuver and field artillery commanders by restricting movement, observation, fields of fire, communications, and target acquisition.

Jungle battles are characterized by limited visibility and close combat. The preponderance of fire support is employed danger-close, and must be carefully coordinated with supporting fire direction centers (FDC) and adjacent maneuver units. Adjustment is frequently conducted using creeping techniques and sometimes by sound.

The following factors must be considered when employing field artillery in a jungle environment.

MUNITIONS EFFECTS

Heavy vegetation degrades the effects of all types of munitions. More firepower or alternative fuze action is normally required to achieve effective results.

The following guidance is offered when requesting or selecting shell-fuze combinations:

- **Proximity (VT) fuzed rounds lose much of their effect in tree canopies or other thickly vegetated areas by exploding above or high in the treetops.**

- **Delay fuzes give better effects in heavy vegetation.**

- **Time fuzed rounds also tend to lose some effect, in the jungle by exploding in tree canopies, due to the fuze impact function. Time fuzes, however, can be extremely effective when used for close, defensive fires (300 to 900 meters) with high-explosive (HE) munitions.**

- **Superquick fuzing is useful in open, lightly vegetated, or secondary growth areas. Rounds may explode too soon in high, thick tree canopies, and like time fuzes, lose most of the shell's effect. In low canopy jungles, however, airbursts can increase casualty-causing effects. As the tree canopy is thinned out by bursting munitions, a switch to VT or time fuze maybe required if airbursts are the desired method of attack. Also, superquick fuzes should not be used in muddy or water covered terrain, such as swamps or paddies. The effects of superquick fuzed HE ammunition will be sharply reduced as blast and shell fragments will be directed upward.**

- **Delay fuze can penetrate thick canopies. The fuze triggers in the treetops, and detonates the round in the air at a lower level. However, when used against troops in inundated areas or areas with soft, boggy soil conditions and no canopy to trigger the fuze overhead, delay rounds will bury themselves in the ground, losing their antipersonnel effect.**

- **Improved conventional munitions (ICM) should not be used in thick foliage and water-filled rice paddies. The ICM submunitions may hang up in heavy foliage or fall undetonated to the ground, creating a hazardous area, or they may submerge in water, rendering them ineffective. Appropriately fuzed HE rounds should be substituted for ICM, as necessary.**

Rearward protection for individual fighting positions must be provided when manning a perimeter around an artillery unit (105-mm) which has beehive ammunition. Infantrymen must also be advised of the signals for firing antipersonnel (beehive) munitions and periodically see a demonstration of effects to insure understanding and compliance.

The heat and humidity of tropical areas speeds the deterioration of all types of ammunition. Further, ammunition packing materials and subcomponents are also subject to more rapid deterioration and require protection.

This protection should include:

- **Adequate shelter and ventilation for ammunition.**
- **Rotating stocks. The first-in should be the first fired.**

OBSERVATION

Ground observation is limited, and aerial observation may be more advantageous. All available air assets, proficient in observed fire procedures, to include USAF, should be used when priorities and level of risk/advantage are favorable. Limited visibility might also make it difficult for ground observers to locate themselves and other friendly units. Determining location by one of the indirect fire means may be necessary. Marking rounds may be fired. In severe cases adjustment is possible by sound.

POSITIONING

Positioning of artillery pieces may pose problems due to soft soil, lack of overhead clearance, and piece masking by trees and terrain. When operating in unfamiliar areas, the battery commander's reconnaissance is important. Construction of prefabricated firing platforms, in accordance with FM 5-15, may be necessary when operating in low-lying areas or near bodies of water. In an area where no suitable firing positions are known to exist, an engineer representative should accompany the battery reconnaissance party. The engineer can provide invaluable expertise when a firing position must be carved out of the jungle or selected on poor terrain.

MOBILITY

The mobility of field artillery units may be impaired by bad weather, lack of roads, or poor terrain trafficability conditions. If roads are used for movement, proper preparations must be made. Roadsides may lack adequate room for firing positions during road movements. Road movement requires ground security forces, supporting artillery, and helicopter or USAF cover to maintain clear routes and provide protection and fire support. If roads are too restrictive, artillery units must be well trained in airmobile operations as this may become their primary mode of transportation in a jungle environment. Engineer support may be required to clear and maintain roads and to prepare adequate LZs.

SURVEY

Line-of-sight limitations severely hamper survey operations. Target area survey may be virtually impossible. Radar, celestial observation, simultaneous observation, hasty survey, and observed firing charts should be used to help offset the probable lack of survey. Firing positions should receive the highest priority in survey planning.

COMMUNICATION

Radio relays and elevated antennas are often required to overcome line-of-sight communications restrictions. Cross-country wire can be laid by helicopter, but specially trained and equipped crews are required.

PLANNING AND COORDINATION

Jungle battles are normally very decentralized, but fire support planning and coordination must be centralized.

Since movement is difficult, fire support may be restricted by the inability to quickly move mortars or artillery so that they can provide continuous coverage. Advance planning and continuous situation monitoring are necessary to avoid such difficulties.

Observed fires are coordinated routinely. Unobserved fires require particular attention to protect friendly units because of the observer's difficulty in accurately locating himself and targets. All indirect fires must be coordinated by fire support officers at all levels. Fire support officers must constantly monitor the location of supported maneuver units to be capable of providing rapid clearance when required.

BATTERY DEFENSE

When the enemy situation is vague or significant enemy guerrilla activity is present, artillery units can be collocated with or provided security by elements of an infantry battalion. Additional security is needed because of the artillery's limited capability for self-defense in the face of a determined enemy attack. The artillery unit's capabilities will also enhance the maneuver unit's defense. When the infantry unit provides security, this permits the battery to fulfill its primary mission of fire support.

The star formation of howitzers is optimal for all-round defensive coverage. However, due to terrain restrictions, a lazy W or variation thereof may be utilized.

Additional security in the form of fire support must be planned for and provided by adjacent artillery and mortar units. This additional fire support should be planned on and around unit positions with both HE and illumination. Once in position, the battery's howitzers, FDC, and TOC form an inner perimeter; the infantry occupies the outer perimeter. Each howitzer will be assigned a sector for direct fire. Each howitzer section chief will make a range card. It will include data necessary to engage targets within his assigned sector.

Each howitzer should have beehive (105-mm only) and/or **HE**, with point detonating and time fuzes allocated for battery defense. Battery defense can be improved by collocating one 81-mm mortar with the battery. The mortar can provide close-in illumination and close indirect fire (within 70 meters of its own position). As soon as possible, howitzer positions should be parapeted, vital installations dug in, and fighting and sleeping positions prepared.

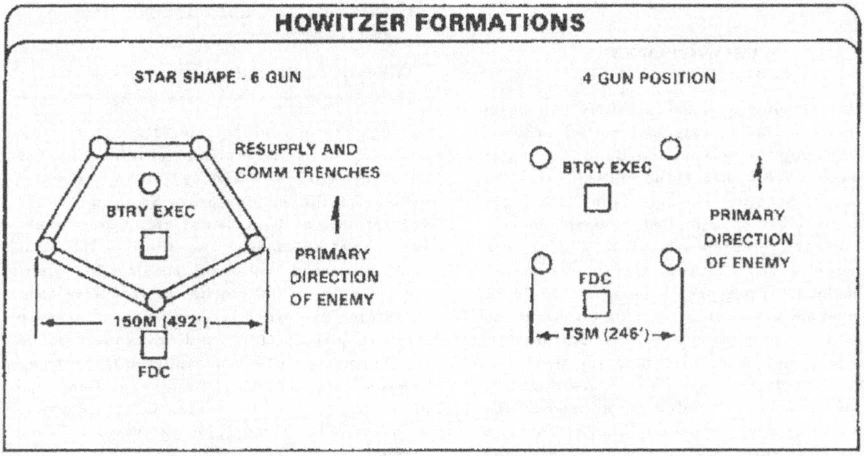

SPLIT BATTERY OR PLATOON OPERATIONS

In order to provide artillery coverage to separated maneuver elements, the battery may have to operate in a split battery or in platoon configurations. If split, the firing elements should be located within maximum range so that both elements can mass fires on the most important targets. Each firing position should be laid out so that each gun can support another gun's position with direct fire, and so located that other artillery units can provide supporting indirect defensive fires. If firing positions are too widely separated, mutual support and the massing of fires will not be possible. Fractional battery operations have a major disadvantage of reducing the number of battery personnel who can be used for battery defense. Augmentation with at least an infantry platoon at each location becomes essential. Separated firing elements should be collocated with a maneuver battalion's defensive position. The battery commander will probably go where the greatest threat exists or with the larger part of the battery. The executive officer will command the other elements of the battery. The communications assets of the battery will be distributed so that each element can monitor and transmit on at least two frequencies. For sustained fractional battery operations, the FDC section must be augmented with additional personnel. This can be done when battery personnel are cross-trained. Split techniques should be avoided where possible as they violate the critical principle of mass and result in drastic reductions in effectiveness.

STRONGPOINT

If it is anticipated that a battery will be in one position for an extended period of time, the position should be hardened to offer protection from direct and indirect fire. In order to prepare the position, adequate materials must be provided, and the battery commander must work closely with the engineers and the maneuver unit commander. If possible, a position should be selected near a usable road to facilitate the movement of heavy engineer equipment, supplies, and ammunition to the position. However, if the strongpoint is accessible only by air, the time needed to construct the strongpoint will be greatly increased. The initial construction efforts will be directed toward the construction of a defensible tactical position by nightfall of the first day.

The following must be completed first:

- **Fighting positions with overhead cover**
- **Howitzer positions with parapets and overhead cover for ammunition**
- **Tactical wire emplaced and Claymore mines positioned**
- **FDC and TOC positions dug in with overhead cover**
- **Range cards and defense plans made**

The firing unit should occupy the position in a formation which provides 6400-mil coverage and optimal battery defense. As the perimeter is expanded, any vegetation which presents site-to-crest problems must be cleared. As soon as the perimeter has been improved, construction efforts are centered on the howitzers. Howitzer positions are reinforced with trail logs, erosion control, powder pits, and revetment of the inside and top of the parapet, using logs or ammunition boxes and sandbags. Sandbags alone can be used to parapet the howitzer, in which case, 1,500 to 2,000 sandbags for each parapet are needed.

105-MM HOWITZER POSITION DIMENSIONS

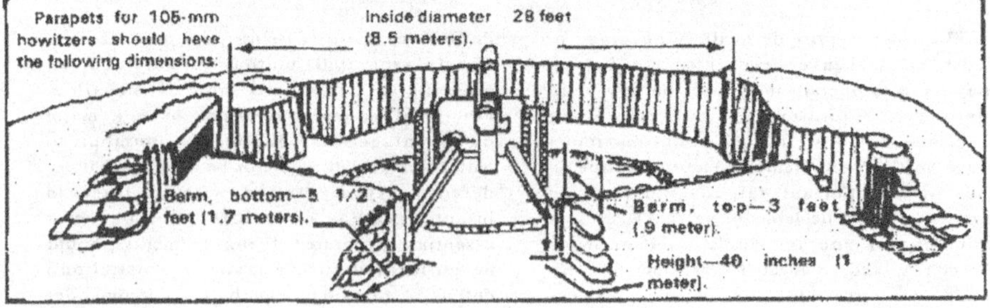

Larger caliber weapons will require special assistance from supporting engineers.

NAVAL GUNFIRE

Naval gunfire, along with close air support, is employed in amphibious operations prior to the landing of artillery units. After maneuver and artillery elements are established ashore, US Navy ships can provide reinforcing artillery support in jungle areas near the coast. Destroyers or smaller inshore fire support ships may be able to provide support along large jungle rivers.

When a unit is conducting operations within range of naval gunfire, a naval gunfire liaison officer will normally be located in the nearest fire support coordination center. He will arrange for ground and aerial observers for the maneuver units. If these are not available, forward observers can adjust the ship's fires. The call-for-fire and adjustment procedures for naval gunfire are basically the same as for artillery support. *(See FM 6-40 for details.)*

NOTE: Naval gunfire provides low trajectory ordnance. This might be a problem in heavily forested jungle areas because the projectile may strike tall trees, detonating prematurely over friendly troops.

TACTICAL AIR SUPPORT

The roles of TACAIR forces are the same in the jungle as elsewhere. In the jungle, where contacts are often unplanned, these aircraft offer another source of responsive combat support. Heavily armed fighter-bombers may be able to engage targets closer to friendly troops than can artillery.

FORWARD AIR CONTROLLERS

Airborne forward air controllers (FAC) are more effective than ground FACs in densely vegetated areas because they have a broader view of the battlefield. The airborne FAC usually can see the target area better for controlling aerial delivery of munitions. All means must be used to mark friendly and enemy locations—smoke, mirrors, and panels. Fires already being delivered can also be used to mark targets. As is true with other aerial systems, the air defense threat is a primary consideration when using airborne FACs.

MUNITIONS

The munitions used vary with the type vegetation and terrain. Unfinned napalm and cluster bomb units may lose much of their effect if they detonate in the thick canopy of primary jungle. Finned napalm, on the other hand, is more stable and accurate and may be used as close as 160 meters to friendly troops in more open areas. The "hard bombs" range in size from 250 to 3,000 pounds and can be used in primary jungle. The FAC will advise danger-close distances. Another weapon, the 20-mm cannon, can be brought in as close as 25 meters from troops under cover or 200 meters from exposed troops.

AIRCRAFT

The A-10 aircraft is particularly well suited to operate in the jungle. It flies slow enough so that the pilot can see his targets. It can also remain in the area for a long time. The A-10 can employ any of the Air Force's close air support munitions, and is fitted with the 30-mm cannon, which is especially well suited for ground support. Safety considerations for the 30-mm cannon are the same as for the 20-mm cannon.

When requesting TACAIR, the requester should specify the type terrain and vegetation it will be targeted on. This will help the Air Force decide what type of ordnance and fuzing to employ.

AIR DEFENSE AND AIR DEFENSE ARTILLERY

Jungles normally provide good concealment from the air. This may reduce air defense requirements. Thickly vegetated areas increase the effectiveness of passive air defense measures, such as camouflage and dispersion. Exposed areas, such as clearings, roads, or river crossings, and facilities where troops tend to concentrate, such as supply points and headquarters, will require air defense artillery (ADA) protection.

The jungle affects ADA mobility, maintenance, and communications in the same way as it affects other combat support resources. In addition, target acquisition is more difficult, and radars are less effective. ADA gunners may not be able to see enemy aircraft until they fly over their position. Reaction times will be short, and many engagements will take place after crossover.

PRIORITY FOR AIR DEFENSE

The jungle concealment afforded to maneuver forces decreases the likelihood that enemy air will attack them. Thus, it is more likely that air attacks will be directed against base complexes, airfields, and logistical facilities. These areas will normally receive first priority for air defense in jungle operations.

HAWKS

Hawk battalions will normally be assigned general support missions for the defense of critical assets, such as base complexes, airfields, port facilities, and lines of communications. As much as possible. Hawk coverage will be extended over the areas in which maneuver forces are operating. To guard against attack by an infiltrating enemy, Hawk elements may be located within base complexes. While this reduces the security problems, it may sacrifice some of the Hawk's early engagement capability. Radars may have to be mounted on towers or berms to be effective.

CHAPARRAL/VULCAN

Chaparral/Vulcan battalions in the jungle can be used in a similar manner to Hawk battalions. In addition, Vulcans can be used to protect convoys from both air and ground attack. Chaparrals may be prepositioned to protect traffic at critical points. When used in this manner, Chaparral/Vulcan units in the jungle are normally attached to the maneuver units they support. Chaparral/Vulcan positions which afford 360-degree coverage and long-range fields-of-fire will be hard to find. Chaparral/Vulcan units may have to clear trees and shrubs to have adequate firing positions.

REDEYE AND STINGER

Redeye and Stinger will cover maneuver units when they are crossing open areas, such as rice paddies or rivers. Redeye and Stinger gunners will also find it hard to find positions which offer 360-degree coverage and long-range fields of fire, and may also have to clear trees and shrubs for adequate firing positions. These positions should only be occupied long enough to fire because they are easily detected from the air. Redeyes and Stingers defending convoys on jungle roads should be positioned within and move with the convoys. Redeyes and Stingers in jungle operations are normally attached to the units they support.

ENGINEERS

Engineers "open up" the jungle. Road building and repair; installation, port, and airfield building; and river crossing operations are continuous tasks for the engineers.

Like most other type supporting units, engineers do their jobs much the same in the jungle as anywhere else. But, also like other type units, they have special considerations to deal with.

ROADS

Road construction in the jungle is usually affected by (1) poor drainage and heavy rainfall, and (2) poor subgrade foundation.

These problems can be dealt with by:

- **Avoiding low ground for road construction**
- **Laying long sections of pontoon bridging, corduroy, or chespaling road through low swampy ground**
- **making roads wider, which thins the overhead foliage, so the sun can get through to dry them out**
- **Using subgrade materials to support heavy traffic** (See TM 5-330 for details.)

AIRFLELDS

Construction of LZs, helipads, and airstrips in remote areas are also important engineer tasks in the jungle. To accomplish these tasks, engineers are inserted into the area to be cleared. Using demolitions and tools, they make an initial clearance of trees and underbrush. For large jobs, follow-on engineers with heavy equipment are brought in to finish the work.

THE LAYING OF A CHESPALING ROAD

Another problem is protecting the surface of these sites from erosion in the rainy season, and protecting aircraft from dust in the dry season. Steel matting, T17 membrane (a tough rubberized fabric), or Peneprime (oil surfacing) are all materials which will keep the dust down on jungle LZs.

PRIMARY SUPPORT

Engineers support the infantry in the jungle primarily by clearing the way for the movement of friendly forces (mobility) while impeding the movement of the enemy (countermobility). Both of these types of support enhance the third area of engineer support—survivability.

Some specific ways engineers can help in all three areas are:

- **Mobility. Improving or building lines of communications and helping move troops and supplies across barriers—particularly water obstacles.**
- **Countermobility. Creating obstacles to support the defense; removing them to assist in the offense.**
- **Survivability. Constructing strongpoints and assisting in the construction of other defensive positions.**

It is also essential that the infantry support the engineers. Engineers are extremely vulnerable during construction activities and need infantry protection to do their jobs.

COMBAT SUPPORT COMPANY IN A JUNGLE ENVIRONMENT

The infantry battalion has its own direct and indirect fire support, reconnaissance/ security, and communications section. Their effective employment is critical to the commander's ability to mass his combat power against the enemy. All of the battalion's combat support elements except the communications platoon are assigned to the combat support company.

The combat support company commander has a unique role in that he:

- **Is a special staff officer, making recommendations on the employment of his elements to the battalion commander**
- **Retains command responsibility for the platoons and sections of the company even though he normally loses operational control in a tactical environment (He still supports them logistically and administratively unless the platoons or sections are attached to a rifle company.)**
- **May become the battalion combat base commander** (See battalion combat base, chapter 5.)
- **May be used as a maneuver element commander**

The battalion commander must determine how to use the combat support company elements through a thorough examination of the battalion's mission and the combat support company's state of training. Only then can he select the appropriate role for the combat support company commander and his unit.

If and when the combat support company commander becomes a maneuver element commander, the commander of headquarters and headquarters company must assume the responsibility of commanding the combat base. In addition, soldiers must be taken from the least essential jobs in the battalion trains to replace the combat support company soldiers. Normally, the scout platoon, the antitank platoon, and the battalion reserve element would be formed into a maneuver element.

FUNCTIONS OF THE SCOUT PLATOON

The mission of the scout platoon is to perform reconnaissance, provide limited security, and assist in controlling the battalion's movement. The scout platoon is normally employed as a unit under battalion control. However, under certain conditions it may be attached to another unit or operate as a separate element with an area of operations of its own. Generally, under all three of these conditions, it is the primary mission of the scout platoon to find the enemy. In a jungle environment it normally performs route, zone, and area reconnaissance. The scout platoon has a limited capability to conduct security operations, to provide early warning of enemy maneuver, and to deny the enemy information concerning the battalion disposition and movements. The primary security mission for the scout platoon is the screen. A screen is a series of OPs from which enemy movement can be observed. When vegetation allows good observation, OPs are located on high ground and patrols are only used for local security around the OP. When observation ranges are very short, the OPs may have to be moved to the lower areas. Patrols between OPs are required to detect enemy movement. The distances between OPs may be extremely long so they must be alert for signs of enemy infiltration.

FUNCTIONS OF THE ANTITANK PLATOON

The primary mission of the antitank platoon is to destroy enemy tanks. Its secondary missions are to engage other vehicles, crew-served weapons, fortified positions, and other point-type targets. Against an enemy with no tanks, or in terrain which does not permit the use of long-range antitank weapons, the antitank platoon may be used as a rifle platoon.

CHAPTER 7

Combat Service Support

Section I. GENERAL

Combat service support elements in a jungle environment retain the same basic missions and capabilities as in other environments. However, they must make adjustments due to terrain, weather, and vegetation.

Jungle operations subject personnel and equipment to effects not found in other environments. Trafficability and security problems, discussed in other chapters, often affect combat service support elements as much as maneuver forces.

The lack of an extensive all-weather transportation network in many jungle areas makes the mission of support units more difficult. Transportation difficulties may dictate that maneuver units be resupplied by air, pack animals, or human portage.

US Army forces in the jungle will normally operate at the end of a long line of communication. Cargo space must not be wasted on unessential items. Transportation priority must be given to essential materials, and support areas must not become crowded with comfort items.

Jungle combat operations are characterized by ambushes and infiltration. The security threat caused by infiltrators will require that lines of communication be patrolled frequently and convoys be escorted. Consequently, it is essential that combat service support is performed as far forward as the tactical situation permits. This not only improves response time, but reduces road movement and allows the combat service support elements to take advantage of the security offered by combat units.

The heat and humidity ofjungle areas also have harmful effects on men and equipment.

CONTENTS

Weapons tend to rust quickly, and must be cleaned and oiled more frequently than in most other areas.

Canvas items rot and rubber deteriorates much faster than in more temperate areas.

Battery life is shorter than is normal.

Electrical connections corrode quickly.

Lenses and dials become quickly fogged with internal moisture.

Troops drink more water, requiring greater water purification and transport means.

Section II. LOGISTICS

TRAINS

The variations ot terrain, weather, and vegetation in the jungle affect the organizing, positioning, and securing of combat service support.

EFFECTS OF THE JUNGLE ON OPERATING THE TRAINS

Organizing the trains is often simplified in the jungle. Because of the terrain, aerial resupply will usually be common practice. The responsiveness provided by aerial resupply requires fewer supplies stockpiled in the combat trains. Thus, combat trains consisting of medics, a maintenance element, and small amounts of C rations, ammunition, and lubricants may be the only requirement. The remainder of the logistical assets will be located in the field trains, to include: petroleum, oils, and lubricants (POL); most vehicles; stockpiled ammunition and other ordnance items; the aid station; rations; maintenance contact teams; and, when available, water purification facilities.

Trains should be far enough forward to respond quickly to troop needs, yet far enough to the rear to have minimum exposure to the enemy. The jungle offers excellent sites for concealment, defilade, and dispersion. Since most resupply is done by air, the combat trains may often be located with the field trains in the brigade trains area. Thus, they can be a greater distance to the rear than normal. All trains should be located near LZs. Field,

A JUNGLE LANDING ZONE

combat, and company trains should also be located close to a road, river, or trail which may be used as an alternate method of resupply. If none of these are available, LZs and supply routes may have to be constructed.

Securing the trains area is a major problem in the jungle. Infiltration of enemy units into rear areas is highly probable. In the defense, the trains should be located within a reserve unit defensive perimeter *(see IV, chap 5)*. In the offense, the trains personnel normally have to provide their own security. If other forces are available—for example, the reserve—they are integrated into this security plan.

Frequent displacement of the trains is often required, either to evade enemy infiltration or to keep up with combat elements. The combat trains usually move with the combat elements when they displace. They can move independently of the forward combat elements if a unit is provided to secure them. The field trains will normally displace when the brigade trains displace.

RESUPPLYING UNITS

Airlift is key to jungle resupply operations to include movement of supplies to attacking or defending units. Returning aircraft should be used to evacuate casualties.

Since the combat trains in a jungle environment are smaller yet less mobile than in more open terrain, most of the class III (POL) and V (ammunition) will be stocked in the field trains. The helicopter can provide timely resupply when ammunition, POL, etc., are not stored in the combat trains. These supplies can often be moved more quickly by air (helicopter or airdropped by parachute) from the field trains than overland from combat trains. The normal method in the jungle is to deliver supplies directly to forward companies (unit distribution).

SECURITY OF SUPPLY ROUTES

A special security problem exists when supplies are not moved by air. Enemy ambushes and mines are a constant threat on ground supply routes. The following actions may be taken to minimize the threat to supply routes.

Commanders should locate fire support elements, command posts, and defensive positions near supply routes. Subordinate units can then conduct clearing and route security operations. Each unit should be given an area to clear and secure. Maneuver units should operate on the flanks of the route to guard against ambush.

CLEARING TEAM

Infantry, armor, and engineer elements can organize mine-clearing teams which methodically search for and clear mines, bodytraps, and roadblocks. Engineers with mine detectors and probes sweep the road. The infantry and tanks provide security against ambush. Any mines detected are marked and either deactivated or blown in place.

Working behind the mine-clearing teams, dump trucks fill holes in the road with gravel. Other engineer teams replace culverts, repair bridges and stream crossings, and clear vegetation back from the edge of the road.

The road-clearing process is repeated each morning before traffic starts to move. The enemy may replace removed mines during the night.

Security against ambush and attack is provided mainly by patrols to either flank of the route to discover enemy forces before they can get into position to launch an ambush. These patrols should range as far out as possible to reduce the danger of the main body being ambushed. Routes should be patrolled at irregular intervals throughout the day and night.

Clearing the vegetation back from the road's edge will also help to prevent ambush, but may be a very large undertaking. It also increases the enemy's difficulty of mining the road, and facilitates friendly aerial observation and airmobile operations along the road.

Airborne artillery observers, forward air controllers, attack helicopters, and air cavalry reconnaissance increase the security by flying reconnaissance and surveillance missions.

At night, counterambush patrols should be sent out. Scout dogs are very useful to these patrols in detecting the enemy. As additional security on the supply route, STANO devices should be emplaced to detect enemy movement at critical points along the route.

Bridges are vulnerable links in any supply route. They must be protected. A static security post may be required to secure the bridge against attack by guerrillas or infiltrators. Static security posts are best organized in a perimeter defense. The size of the defending force depends on the mission, the size and characteristics of the hostile force, and the importance of the bridge to the mission. Such posts in remote areas will be larger than those nearer supporting forces. The parent unit must be prepared to counterattack with reserves to assist each post.

It will usually be necessary to form armed convoys, escorted by armed vehicles from a military police, cavalry, or maneuver unit. At least one helicopter should fly ahead of the main body to give early warning of ambushes or evidence of hasty mining and provide additional communications capability. On-call fires should be planned on likely ambush sites. Convoys should not be scheduled at regular intervals.

If ambushed, the escort vehicles should immediately return fire. The cargo vehicles should be moved out of the kill zone as rapidly as possible. Attack helicopters and trailing combat vehicles can then maneuver to destroy the ambush.

CLASSES OF SUPPLY

CLASS I (SUBSISTENCE ITEMS)

Nonperishable canned, dried, or dehydrated items primarily should be used when the unit is not actively engaged in combat. This requires kitchen facilities, except for refrigeration. Rations (individual combat meal, long-range patrol, or small detachment rations) normally will be issued to units actively engaged in combat. The number of rations earned by the individual soldier should be determined by such factors as weight, how and in what quantities food can be brought forward, when resupply will be conducted, and the estimated duration of the operation. Hot meals should be served when the combat situation allows. These should be prepared in the field trains and delivered by helicopter. Feeding is usually accomplished during the day because of the danger and difficulty of movement at night and the possibility of enemy ambush or night attack.

CLASS II (INDIVIDUAL EQUIPMENT AND GENERAL SUPPLIES)

In a tropical environment, rapid deterioration is a primary consideration in class II supply. Issue clothing, particularly combat boots and socks, lasts a very short time. Normal wear and tear on clothing may require a resupply every 5 to 6 days. Class II resupply can be a major morale builder. Class II requirements should be estimated well in advance and special provisions should be made for timely resupply. At battalion level, limited emergency supplies of assorted combat boots, socks, uniforms, and similar items of short wear periods should be stocked in the field trains.

CLASS III (POL)

The supply of class III items does not initially present a great problem, as relatively few vehicles will be forward during the operation. However, aviation fuel will be required in large amounts. Units must establish class III distribution points for vehicles and aircraft, as required, in the field trains. Tank trucks will be used as far forward as possible. Collapsible fuel drums can be brought forward by helicopter. Units must take care to insure that diesel and motor gasoline fuel drums are available and marked. When circumstances permit the use of tracked vehicles, planning must include additional POL since tracked vehicles normally have a high fuel consumption rate when operating in rugged terrain. Additional lubricants for weapons will also be required due to the greater requirement for cleaning and protection.

CLASS IV (CONSTRUCTION MATERIALS)

Barrier material and special equipment is extremely bulky and heavy. The use of large amounts of it for defensive positions will create transportation problems. Heavy lift helicopters are the most practical means to move these items.

CLASS V (AMMUNITION)

Because of weight and bulk involved, resupply of ammunition and explosives presents a difficult problem. The nature of the terrain and climate in the jungle magnifies the problems in ammunition storage. Steep terrain should be avoided when selecting suitable sites for storage. Level, firm ground may become too soft in heavy rain. Rain and intense heat combine to speed the deterioration of containers and ammunition, especially if it is taken out of the packing material. To counteract these conditions, adequate dunnage, shelter, and ventilation must be used. Ammunition in the field trains should be left in packing materials until ready for use. Frequent checks should be made for deterioration of packing material.

When the ammunition is unpacked and taken to the field, it is more susceptible to rain, heat, and humidity. Therefore, as a general rule, ammunition which is not used within two weeks should be exchanged for new ammunition. While in the field, however, measures must be taken so that ammunition is never exposed needlessly to the weather. It should be carried in ammunition pouches or containers and protected when stored in positions.

CLASS VT (PERSONAL DEMAND ITEMS)

Such items as soap, cigarettes, candy, etc., are important morale builders. They should be supplied on a regular basis.

CLASS VII (MAJOR END ITEMS)

Storage areas must be designed to protect these items from the jungle environment. Weapons, vehicles, and other items in storage must be inspected more frequently than in other environments. Protective lubricants must be maintained on all surfaces that can rust.

Operational readiness of all stored equipment must be insured by frequent inspections and maintenance. Commanders should turn in those items for which they have no need in the jungle. When major end items are not in use, they are stored or maintained in rear areas.

CLASS VIII (MEDICAL SUPPLIES)

Because the heat and high humidity can cause deterioration of certain medical materials and medicines (for example, penicillin, hydrocortisone liquids, sulfa liquids), frequent resupply and replacement of these items may be required. There will also be a priority need for refrigeration equipment to store class VIII supplies. There will be an increased requirement for antiseptic material because the high heat and humidity increase the incidence of infection in wounds.

CLASS LX (REPAIR PARTS)

Equipment parts that deteriorate or wear out faster in the jungle environment must be determined. Prescribed load list (PLL) must reflect the increased turnover of these parts.

CLASS X (NONMILITARY ITEMS)

These items, for example, agricultural tools and equipment, are not required for the conduct of conventional combat operations in the jungle; however, they are very important in the conduct of civil-military operations. These items should not be stocked unless a unit is actively participating in a civil-military operation.

OTHER SUPPLIES

Maps exposed to humid jungle air quickly deteriorate. A greater stockage of maps will be required than in other environments.

Water is critical in the jungle. Water must be obtained from the divisional water points using battalion transportation or purified by units themselves in the field. There is a greater demand for water purification tablets and water points. Units should also carry a greater number of lyster bags and other water containers. Close supervision should be exercised to insure that soldiers do not drink unpurified water.

The lack of suitable landing zones makes water resupply extremely difficult if the aircraft cannot land. Numerous techniques have been tried in dropping water through the jungle canopy, for example, free drop, parachute. However, they have met with only limited success and for the most part are impractical. The following water resupply technique has proven to be extremely successful. Water should be placed in either collapsible 5-gallon containers or in 2-quart containers. These containers should be placed inside lightweight, portable containers, that is, rucksacks or duffel bags, which a soldier can quickly move off the DZ. No more than 10 gallons of water (80 pounds) should be placed in each bag for lowering. The bags will be lowered from a helicopter by use of the equipment belay system.

Equipment needed for platoon-size element to supply water by helicopter, using collapsible containers and the Equipment Belay System:

Two 120-foot climbing ropes.

Eleven snap links.

Three 12-foot utility ropes (sling ropes).

One doughnut ring or similar anchoring device.

One rappel D-bag.

Eight 5-gallon containers and four duffel bags.

MAINTENANCE

Maintenance organizations function essentially the same as in other operations. The high humidity and temperature prevalent in jungle areas will increase maintenance requirements. Preventive maintenance on any item affected by moisture and heat is extremely important. Emphasis must be placed on on-site maintenance and the use of aircraft to transport maintenance contact teams and repair parts to unit level. To respond to the need for responsive maintenance support, the number of repair parts for immediate direct exchange must be increased.

TRANSPORTATION

Units fighting in the jungle should consider all types of transportation. Surface transportation facilities are poor in most jungle areas and cannot handle heavy military traffic without extensive improvements. An air line of communication can eliminate many of the problems associated with surface movement.

HUMAN PORTAGE

This is a basic means of moving supplies and equipment in jungle operations. At best, this method is slow, laborious, and inefficient. Yet, in certain extreme situations, it may be the only method available.

WHEELED AND TRACKED VEHICLES

Wheeled vehicles are normally restricted to roads and wider trails, and even these may prove impassable during heavy rains. Sometimes, goods must be transported by cross loading from wheeled to tracked vehicles. For example, large wheeled vehicles move the supplies as far forward as possible, where they are transloaded to tracked vehicles which move them cross-country. In rugged terrain, the supplies may have to be further transloaded to pack animals or native supply bearers.

INLAND WATERWAY

INLAND WATERWAYS

These channels are a very important part of a transport system, especially in remote areas. Large rivers often allow small ships and large boats to penetrate several hundred miles inland. Where smaller streams branoh out, a unit may establish transter points for the transloading of cargo into smaller watercraft. A river that is normally very shallow during the dry season will be deeper during the rainy season, permitting travel by larger craft.

A unit should establish supply points high enough above the level of the water to prevent damage in the event of flash flood.

FLXED-WING TRANSPORTAIRCRAFT

These aircraft can usually operate at greater distances without refueling than cargo helicopters. However, use of fixed-wing aircraft to airland supplies requires more landing strips than may be present. Construction and maintenance of airfields in jungles is a difficult engineer task, but open savanna may be large enough and firm enough to use as an airstrip.

Airdrop of supplies is an alternative to airlanding Airdrop by parachute is a rapid means of delivery and makes deliveries to isolated units possible without further transloading.

Disadvantages include the dispersion of supplies and the possibility of lost cargo in the jungle canopy, vulnerability to local enemy air defense, and requirement for at least local friendly air superiority.

A variation of airdrop by parachute is the low altitude parachute extraction system (LAPES). The parachute is used to pull the load from the rear ramp of an aircraft flying at a reduced speed just a few feet off the ground. LAPES solves dispersion and tree hangup problems, allowing use of the smaller drop zones more common in jungle areas. However, fragile equipment might not survive this type of drop. In areas where the aircraft can make a low-level approach, such as savannas, LAPES may be a valuable resupply technique. It is used with C-130 aircraft only.

LAPES

The container parachute delivery system provides single pass delivery of from 1 to 16 individually rigged A-21 containers into a small drop zone (DZ) at altitudes of **500** to **600 feet.**

When using this system, selection of a DZ is critical to accurate delivery of the bundles. The nature of jungle terrain and the small DZs do not allow for accurate delivery of bundles. The DZ must be at least a 400-meter by 400-meter area. A straight stretch of road or a large field visible from the air may be used. If an open ground area cannot be located, the next step is to try to make the drop in a river, lake, or ocean. If dropped in the water, bundles must be rigged with adequate buoyancy (empty water cans will do). Recovery of the "bundles must be preplanned and executed immediately upon completion of the drop.

Another method which can be used for small loads of some sturdy types of supplies and equipment is the free drop. Here, the load is simply dropped, without parachute, from a low-flying aircraft.

Helicopters are very effective in this technique because they can hover low over a unit and drop the supplies exactly where desired. Because helicopters are vulnerable to small-arms fire, however, the unit on the ground must take adequate security measures. It should establish a perimeter defense around the area over which the helicopter will hover, and it should conduct patrols prior to the resupply to insure that the surrounding area is cleared of enemy.

FREE DROP

MEDICAL OPERATIONS

Medical support in jungle operations follows the basic principles of military medicine, but medical treatment and evacuation will often be complicated by the extended distances and inaccessibility due to terrain and vegetation. The manner in which medical units support tactical operations will depend on the employment of the supported unit and the type of jungle in which they are operating. The relatively high incidence of disability caused by heat, humidity, and insect-borne diseases will require increased emphasis on health and sanitation measures.

The evacuation of wounded in jungle warfare presents numerous problems. Because of the prevalence of rapidly debilitating tropical diseases and the likelihood of quick infection of wounds in the jungle, speedy evacuation is vital. Units must use all available methods to augment organic medical evacuation means. These may include tracked vehicles, pack animals, watercraft, helicopters, and litter bearers. Ground medical evacuation in the jungle will normally be along supply routes, as they are usually given all possible protection from the enemy.

The difficulties of overland medical evacuation in jungle operations emphasize the advantages of air evacuation. When weather, aircraft availability, and friendly air superiority permit their use, helicopters provide fast, comfortabe, and efficient transportation for casualties. The reduction of time between injury and treatment increases the chance of survival and raises morale. A helicopter equipped with a cable hoist system can evacuate casualties from thick jungle vegetation where no landing zones are available.

In some remote and densely foliaged jungles, the only means of evacuation maybe by litter. This is a slow and exhausting task requiring well-conditioned soldiers to carry the litters. At best, litter teams can carry patients only a few hundred meters over rough jungle terrain before needing rest or relief.

There are several types of litters. In very rugged terrain, the mountain-type metal basket litter offers most security to the patient. However, it is bulky to carry even during normal operations. The standard folding litter or field expedient variations made of cloth and poles are less bulky and get the job done just as well if patients are moved with great care.

Litter hauls should be as short as possible. Medical facilities should be as close as practicable to the troops.

PERSONNEL MA NAGEMENT CONSIDERATIONS

REPLACEMENTS

The jungle may appear as a fearsome place to the uninitiated. Individual replacements may doubt their ability to cope with tropical diseases; heat; oppressive humidity; entangling close vegetation; the constant threat of ambush and close combat; wildlife; and periodic isolation from support elements. Commanders must realize that these are natural fears and must provide a period for acclimation and psychological adjustment. All new arrivals must receive a complete orientation on the unit, its mission, and the enemy, and be trained in the routines of jungle living, working, and fighting.

GRAVE REGISTRATION

Direct support for the recovery, identification, and evacuation of deceased personnel will normally be available in the jungle. However, the hot and humid climate presents special problems, and it is important that the remains of the dead be removed rapidly to prevent their deterioration. *This may have profound effect on troop morale.* Hasty burials should be conducted as a last resort. Grave sites should be marked clearly and reported.

APPENDIX A

References

FIELD MANUALS (FM)

3-10		Employment of Chemical Agents
5-15		Field Fortifications
5-20		Camouflage
5-31		Boobytraps (U)
5-36		Route Reconnaissance and Classification
6-20	(HTF)	Fire Support in Combined Arms Operations
6-40		Field Artillery Cannon Gunnery
6-50		The Field Artillery Cannon Battery
7-7	(HTF)	The Mechanized Infantry Platoon and Squad
7-8	(HTF)	The Infantry Platoon and Squad (Infantry, Airborne, Air Assault, Ranger)
7-10	(HTF)	The Infantry Rifle Company
7-20	(HTF)	The Infantry Battalion (Infantry, Airborne, Air Assault, Ranger)
8-35		Evacuation of the Sick and Wounded
19-40		Enemy Prisoners of War, Civilian Internees, and Detained Persons
20-12		Amphibious Embarkation
20-32		Mine/Countermine Operations at the Company Level
20-33		Combat Flame Operations
21-10		Field Hygiene and Sanitation
21-11		First Aid for Soldiers
21-20		Physical Readiness Training
21-26		Map Reading
21-30		Military Symbols

21-31		Topographic Symbols
21-40		NBC (Nuclear, Biological, and Chemical) Defense
21-60		Visual Signals
21-75		Combat Training of the Individual Soldier and Patrolling
21-76		Survival, Evasion, and Escape Manual
23-9		M16A1 Rifle and Rifle Marksmanship
23-90		81-mm Mortar
24-18		Field Radio Techniques
24-20		Field Wire and Field Cable Techniques
24-21		Tactical Multichannel Radio Communications Techniques
29-51		Division Supply and Field Service Operations
30-5		Combat Intelligence
30-10		Military Geographical Intelligence (Terrain)
44-1	(HTF)	US Army Air Defense Artillery Employment
44-3	(HTF)	Air Defense Artillery Employment: Chaparral/Vulcan
57-38		Pathfinder Operations
71-1	(HTF)	Tank and Mechanized Infantry Company Team
71-2	(HTF)	The Tank and Mechanized Infantry Battalion Task Force
90-2	(HTF)	Tactical Deception
90-4	(HTF)	Airmobile Operations
90-6	(HTF)	Mountain Operations
90-10	(HTF)	Military Operations on Urbanized Terrain (MOUT)
90-13	(HTF)	River Crossing Operations
100-5	(HTF)	Operations
100-20		Low Intensity Conflict

TECHNICAL MANUALS (TM)

3-240	Field Behavior of Chemical, Biological, and Radiological Agents
5-277	Bailey Bridge
5-312	Military Fixed Bridges
9-1300-205	Ammunition and Explosive Standards
10-1670-262-12	Operator's and Organizational Maintenance Manual (Including Repair Parts and Special Tool Lists) for Personnel Stabo Extraction System and Cargo and Personnel Lowering Device
11-666	Antennas and Radio Propagation

APPENDIX B

Navigation and Tracking

Section I. GENERAL

This appendix assumes that the reader already knows the basics of map reading and land navigation. For a review of these subjects, the reader should refer to *FM 21-26.*

Section II. JUNGLE NAVIGATION

Navigating in the jungle can be difficult for those troops not accustomed to it. This appendix outlines techniques which have been used successfully in jungle navigation. With training and practice, troops should be able to use these techniques to navigate in even the thickest jungle.

NAVIGATION TOOLS

MAPS

Because of the isolation of many jungles, the rugged ground, and the presence of the canopy, topographic survey is difficult and is done mainly from the air. Therefore, although maps of jungle areas generally depict the larger features (hill, ridges, larger streams, etc.) fairly accurately, some smaller terrain features (gullies, small or intermittent streams, small swamps, etc.), which are actually on the ground, may not appear on the map. Also, many older maps are inaccurate. So, before going into the jungle, commanders and staff should bring their maps up to date.

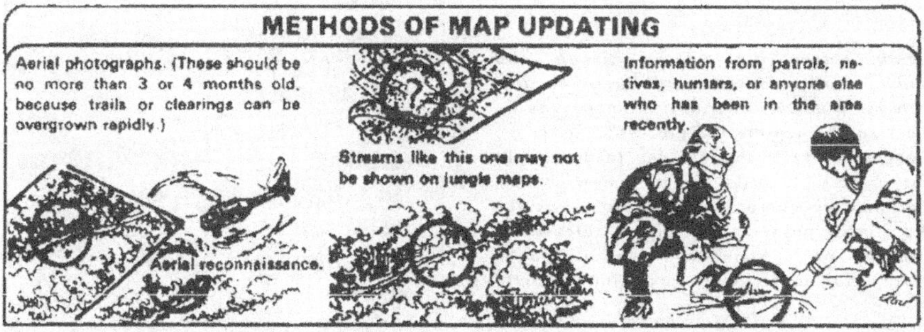

CONTENTS

COMPASS

No one should move in the jungle without a compass. It should be tied to the clothing by a string or bootlace. The three most common methods used to follow the readings of a compass are:

Sighting along the desired azimuth. The compass man notes an object to the front (usually a tree or bush) that is on line with the proper azimuth and moves to that object. *This is not a good method, in the jungle as trees and bushes tend to look very much alike.*

Holding the compass at waist level and walking in the direction of a set azimuth. *This is a good method for the jungle. The* compass man sets the compass for night use with the long luminous line placed over the luminous north arrow and the desired azimuth under the black index line. There is a natural tendency to drift either left or right using this method. Jungle navigators must learn their own tendencies and allow for this drift.

Sighting along the desired azimuth and guiding a man forward until he is on line with the azimuth. The unit then moves to the man and repeats the process. This is the most accurate method to use in the jungle during daylight hours, but it is slow. In this method, the compass man cannot mistake the aiming point and is free to release the compass on its string and use both hands during movement to the next aiming point.

The keys to navigation are maintaining the right direction and knowing the distance traveled. Skill with the compass (acquired through practice) takes care of the first requirement. Ways of knowing the distance traveled include checking natural features with the map, knowing the rate of movement, and pacing.

CHECKING FEATURES

Major recognizable features (hills, rivers, changes in the type of vegetation) should be noted as they are reached and then identified on the map. Jungle navigators must BE CAUTIOUS ABOUT TRAILS—the trail on the ground may not be the one on the map.

RATE OF MOVEMENT

Speed will vary with the physical condition of the troops, the load they carry, the danger of enemy contact, and the type of jungle growth. *The normal error is to overestimate the distance traveled.* The following can be used as a rough guide to the maximum distance covered in 1 hour during daylight.

DAYLIGHT MOVEMENT

TYPE TERRAIN	MAXIMUM DISTANCE (in meters per hour)
TROPICAL RAIN FOREST	1,000
DECIDUOUS FOREST, SECONDARY JUNGLE, TALL GRASS	500
SWAMPS	100 TO 300
RICE PADDIES (WET)	800
RICE PADDIES (DRY)	2,000
PLANTATIONS	2,000
TRAILS	1,500

PACING

In thick jungle, this is the best way of measuring distance. It is the only method which lets the soldier know how far he has traveled. With this information, he can estimate where he is at any given time-something that must be known to call for indirect fire support in a hurry. To be accurate, soldiers must practice pacing over different types of terrain. Each soldier should make a PERSONAL PACE TABLE like this

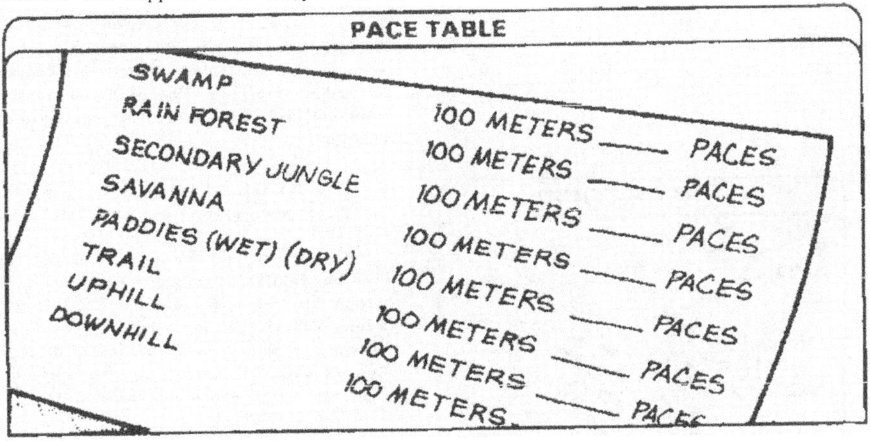

At least two men in each independent group should be compass men, and three or four should be keeping a pace count. The artillery fire support team (FIST) chief should keep an accurate fix on his location, as should the platoon forward observers (FO). He can be a great help to the rifle company commander in matters of navigation. When in doubt, the commander should **do a resection, using artillery marking rounds as outlined in *FM 21-26, paragraph 10.***

NOTE: Jungle foliage will often require that artillery marking rounds be sensed by sound.

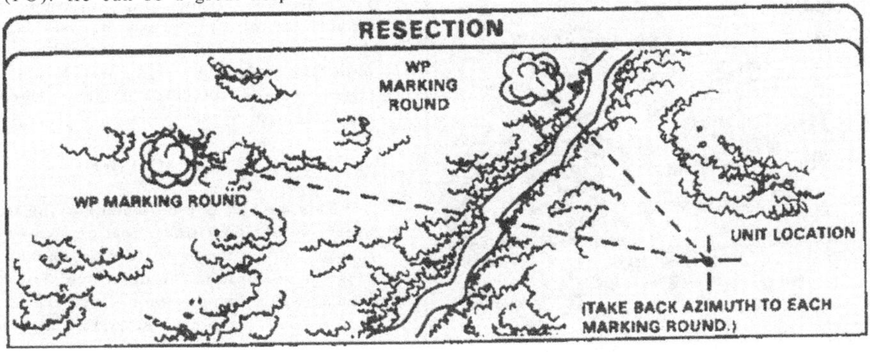

To locate a position by resection, the general location must be known. White phosphorus rounds (airbursts) are then called on two widely separated grids which are not on terrain features like the one the unit is occupying and which are a safe distance from the estimated location. A back azimuth to each of these rounds is taken and plotted on the map. The point where they intersect is the observer's approximate location.

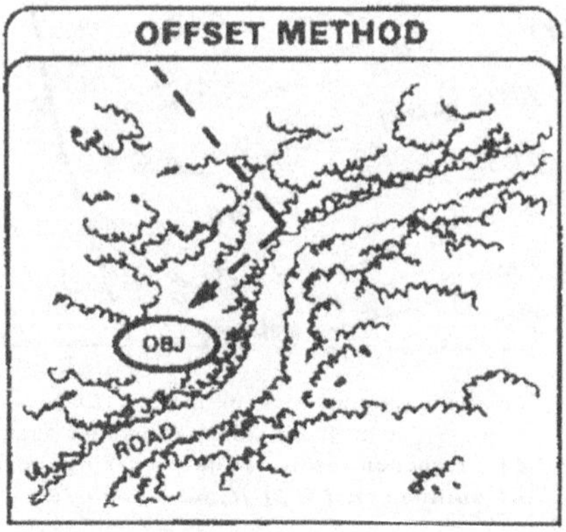

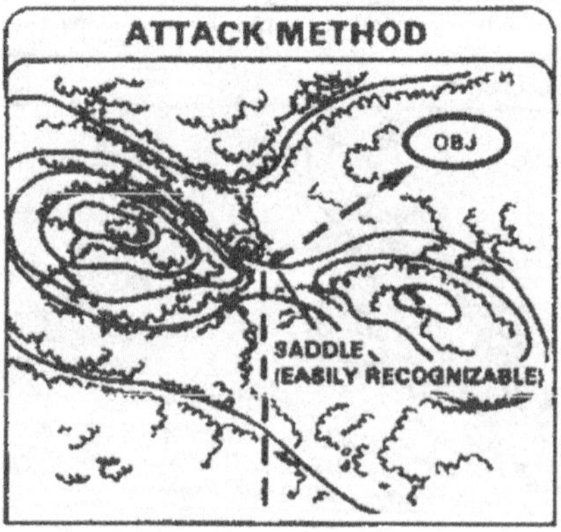

LOCATION OF AN OBJECTIVE

In open terrain, an error in navigation can be easily corrected by orienting on terrain features which are often visible from a long distance. In thick jungle, however, it is possible to be within 50 meters of a terrain feature and still not see it. Here are two methods which can aid in navigation.

OFFSET METHOD

This method is useful in reaching an objective that is not large or not on readily identifiable terrain but is on a linear feature, such as a road, stream, or ridge. The unit plans a route following an azimuth which is a few degrees to the left or right of the objective. The unit then follows the azimuth to that terrain feature. Thus, when the unit reaches the terrain feature, the members know the objective is to their right or left, and the terrain feature provides a point of reference for movement to the objective.

ATTACK METHOD

This method is used when moving to an objective not on a linear feature. An easily recognizable terrain feature is chosen as close as possible to the objective. The unit then moves to that feature. Once there, the unit follows the proper azimuth and moves the estimated distance to get to the objective.

WHAT TO DO IF LOST

Do not panic. Few soldiers have ever been permanently lost in the jungle, although many have taken longer to reach their destination than they should.

Disoriented navigators should try to answer these questions. (If there are other navigators in the group, they all should talk it over.)

What was the last known location?

Did the unit go too far and pass the objective? (They should compare estimates of time and distance traveled.)

Does the terrain look the way it should? (They should compare the surroundings with the map.)

What features in the area will help to fix the unit's location? (They should try to find these features.)

If the unit is unable to locate itself using these techniques, the leader can call for an air or artillery orienting round. However, this may cause a loss of security, particularly if the unit is moving by stealth.

An airspot can usually be obtained from Army or Air Force aircraft. The pilot can be contacted and guided to the general location by radio. A mirror, smoke, panels, or some other signal can be shown to the pilot. He will be able to determine the unit's location and report it to them.

This, also, is a loss of security and should only be a last resort.

NIGHT MOVEMENT

The principles for navigation at night are the same as those for day movement. The problem in night movement is one of control, not navigation. In clear weather, through sparse vegetation and under a bright moon, a unit can move almost as fast by night as by day. If the sky is overcast, vegetation is thick, or there is little or no moon, movement will be slow and hard to control. The following points can assist a unit during night movement.

Attach *luminous tape* to the back of each soldier's headgear. Two strips, side by side, each about the size of a lieutenant's bar, are recommended. *The two strips aid depth perception and reduce the hypnotic effect that one strip can cause.*

When there is no light at all, distance between soldiers should be reduced. When necessary to prevent breaks in contact, each soldier should hold on to the belt or the pack of the man in front of him.

The leading man should carry a long stick to probe for sudden dropoffs or obstacles.

In limited visibility conditions, listening may become more important to security than observing. A unit which hears a strange noise should halt and listen for at least 1 minute. If the noise is repeated or cannot be identified, patrols should be sent out to investigate. Smell, likewise, can be an indication of enemy presence in an area.

All available night vision devices should be used.

NAVIGATIONAL TIPS

- Trust the map and compass, but understand the map's possible shortcomings. Use the compass bezel ring, especially during night navigation.
- Break brush. Do not move on trails or roads.
- Plan the move, and use the plan.
- Do not get frustrated. If in doubt, stop and think back over the route.
- Practice leads to confidence.

Section III. TRACKING

Visual tracking is following the paths of men or animals by the signs they leave, primarily on the ground or vegetation. Scent tracking is following men or animals by the odors they leave.

Practice in tracking is required to achieve and maintain a high standard of skill. Because of the excellent natural concealment the jungle offers, all soldiers should be familiar with the general techniques of visual tracking to enable them to detect the presence of a concealed enemy, to follow the enemy, to locate and avoid mines or booby-traps, and to give early warning of ambush.

Tracking is important in counterguerrills operations where it is often difficult to locate the enemy. Guerrillas who conduct raids and ambushes will normally return to their bases as quickly as possible. Well-developed tracking skills will help units to maintain contact with the enemy.

SIGNS

Men or animals moving through jungle areas leave signs of their passage. Some examples of these signs are listed below.

TRACKING POINTS	
SAVANNA	**ROCKY GROUND**
NOTE: If the grass is high, above 3 feet, trails are easy to follow because the grass is knocked down and normally stays down for several days. If the grass is short, it springs back in a shorter length of time. ■ Grass that is tramped down will point in the direction that the person or animal is traveling. ■ Grass will show a contrast in color with the surrounding undergrowth when pressed down. ■ If the grass is wet with dew, the missing dew will show a trail where a person or an animal has traveled. ■ Mud or soil from boots may appear on some of the grass. ■ If new vegetation is showing through a track, the track is old. ■ In very short grass (12 inches or less) a boot will damage the grass near the ground and a footprint can be found.	■ Small stones and rocks are moved aside or rolled over when walked on. The soil is also disturbed, leaving a distinct variation in color and an impression. If the soil is wet, the underside of the stones will be much darker in color than the top when moved. ■ If the stone is brittle, it will chip and crumble when walked on. A light patch will appear where the stone is broken and the chips normally remain near the broken stone. ■ Stones on a loose or soft surface are pressed into the ground when walked upon. This leaves either a ridge around the edge of the stone where it has forced the dirt out, ora hole where the stone has been pushed below the surface of the ground. ■ Where moss is growing on rocks or stones, a boot or hand will scrape off some of the moss.

<table>
<tr><th colspan="2">TRACKING POINTS CONTINUED</th></tr>
<tr><td rowspan="2">

PRIMARY JUNGLES

NOTE: Within rain forests and deciduous forests, there are many ways to track. This terrain includes undergrowth, live and dead leaves and trees, streams with muddy or sandy banks, and moss on the forest floor and on rocks, which makes tracking easier.

■ Disturbed leaves on the forest floor, when wet, show up a darker color when disturbed.

■ Dead leaves are brittle and will crack or break under pressure of a person walking on them. The same is true of dry twigs.

■ Where the undergrowth is thick, especially on the edges of the forest, green leaves of the bushes that have been pushed aside and twisted will show the underside of the leaf—this side is lighter in color than the upper surface. To find this sort of trail, the tracker must look through the jungle instead of directly at it.

■ Boot impressions may be left on fallen and rotting trees.

■ Marks may be left on the sides of logs lying across the path.

■ Roots running across a path may show signs that something has moved through the area.

■ Broken spiderwebs across a path indicate that something has moved through the area.

</td><td>

SECONDARY JUNGLE

■ Broken branches and twigs.

■ Leaves knocked off bushes and trees.

■ Branches bent in the direction of travel.

■ Footprints.

■ Tunnels made through vegetation.

■ Broken spiderwebs.

■ Pieces of clothing caught on the sharp edges of bushes.

</td></tr>
<tr><td>

RIVERS, STREAMS, MARSHES, AND SWAMPS

■ Footprints on the banks and in shallow water.

■ Mud stirred up and discoloring the water.

■ Rocks splashed with water in a quietly running stream.

■ Water on the ground at a point of exit.

■ Mud on grass or other vegetation near the edge of the water.

</td></tr>
</table>

DECEPTION

The enemy may use any of the following methods to deceive or discourage trackers. They may, at times, mislead an experienced tracker.

These deceptions include:

- Walking backwards. The heel mark tends to be deeper than that of the ball of the foot. The pace is shorter.
- More than one person stepping in the same tracks.
- Walking in streams.
- Splitting up into small groups.
- Walking along fallen trees or stepping from rock to rock.
- Covering tracks with leaves.

WARNING:

A TRACKER SHOULD ALWAYS BE ALERT TO THE POSSIBILITY THAT THE ENEMY IS LEAVING FALSE SIGNS TO LEAD THE UNIT INTO AN AMBUSH.

APPENDIX C

Jungle Obstacles

Section I. GENERAL

Units operating in the jungle will have to cross many obstacles. The most difficult obstacles will be streams and cliffs. In addition, units operating in the jungle will frequently have to insert or extract soldiers and units in places where helicopters cannot land. This appendix covers the skills required to perform these tasks.

Before learning these skills, however, soldiers must be familiar with ropes and knots.

Section II. CHARACTERISTICS AND USE OF EQUIPMENT.

ROPES AND KNOTS

TYPES OF ROPE

Nylon. Nylon rope is most commonly used in climbing and rappelling. The rope is seven-sixteenths of an inch in diameter and is issued in 120-foot lengths. Its dry breaking strength averages 3,840 pounds (plus or minus 5 percent). Strength is reduced by about 20 percent when the rope is wet. It will also stretch about one-third of its length when wet. Nylon sling (utility) ropes are commonly prepared by a unit in 12-foot lengths from older ropes that are no longer used for climbing or rappelling.

Vegetable Fiber. This is readily available in jungle areas as it is made primarily from the fibers of tropical plants.

Manila rope is made from the fibers of the leaves of a banana tree. The lighter the color of the rope, the better the quality. This rope is superior to nylon rope for suspension traverses and rope bridges because it does not stretch as much as nylon, and it is not weakened when wet.

CONTENTS

> The breaking strength and safe load capacity (respectively) for the sizes of manila rope most often used by jungle troops are:
>
> 1 -inch-diameter rope-9,000 pounds/ 2,250 pounds
>
> 1 /2-inch-diameter rope-3,650 pounds/ 660 pounds

Hemp rope is made from the fibers of the hemp plant. This is the strongest of the fiber ropes. It is usually soaked in tar to preserve the rope from damage caused by dampness, but this tar tends to reduce the rope's strength. Also, because of its greater weight, tarred hemp is not practical for use by infantry troops.

NOTE: The breaking strength of a rope is always greater than its safe working capacity. The difference is a "safety factor. " Individual ropes can vary greatly in minimum breaking strength. Even though a rope may not break under this load, the fibers are stretched beyond their elastic limit. Thereafter the strength of the rope is permanently reduced. Exposure, wear, use, and bending decrease a rope's strength over a period of time. This should be allowed for in estimating the strength of a used rope. The strength of a rope that is slung over a hook or contains a knot is reduced by about 30 percent; sharp bends over corners will cut strength by 50 percent; sand or grit between the fibers will quickly cut the fibers, and sharply drop the overall strength of the rope.

CARE OF A ROPE

Clean a muddy rope by washing it in water, but not in salt water.

Do not pull a rope oversharp edges. Place layers of heavy cloth or grass between the rope and any sharp edge to prevent the cutting of fibers.

Do not drag a rope through sand and dirt, or step on it, or drive over it.

Keep a rope dry. If it gets wet, dry it as soon as possible to prevent rotting. (A mildewed rope will have a musty odor and inner fibers will have a dark, stained look.)

Do not leave a rope knotted or tightly stretched any longer than needed.

Never splice a climbing or rappelling rope.

Inspect a rope often, both the outside and the inside. Untwist a few strands at different points to open the rope to check the inside.

Melted nylon and dark streaks indicate burns. Nylon rope burns when it rubs against other nylon ropes. Nylon ropes should never be tied in such away that there is rope-to-rope friction.

Dirt and sawdust-like material inside the rope indicates damage.

A rope should be checked at a number of different places—any weak point in it weakens the entire rope.

Whenever any unsafe conditions are found in a rope, it should be destroyed or cut up in short pieces. This w-ill prevent use of the rope for hoisting. The short pieces can be used for toggle ropes and for other purposes which do not involve load bearing operations.

KNOT-TYING TERMS

Knots to tie the end of a rope to an object (anchor knots).

The **bowline** will not slip under strain, yet it is easily untied.

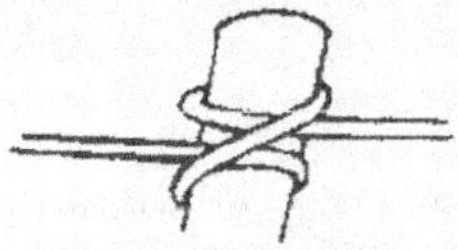

Clove hitch used as an intermediate anchor with tension applied at all times to prevent slipping.

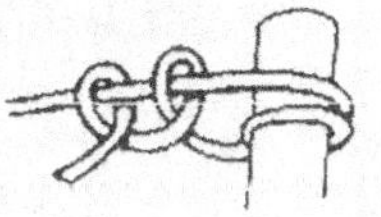

Round turn with two half-hitches.

Knots to tie the ends of two ropes together.

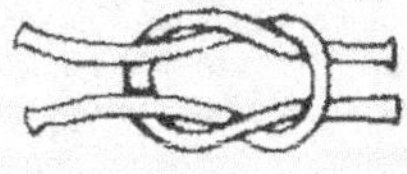

A **square knot** is used to tie ropes of equal diameter together.

The **double sheet bend knot** is used to tie ropes of unequal diameter together.

Middle-of-rope knots.

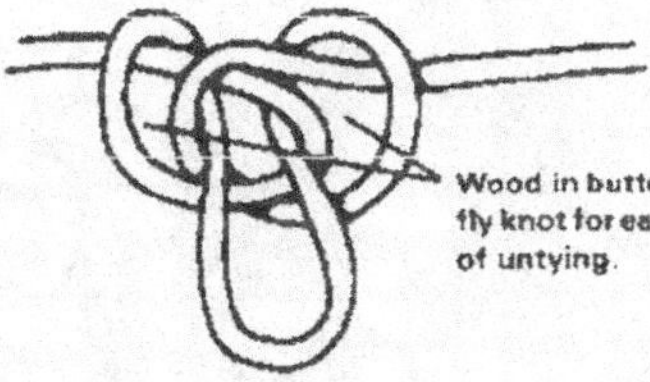

The **butterfly knot** is used to pull a line tight.

NOTE: After a butterfly knot is tied for a rope bridge, or for any other purpose where a great amount of strain or tension is applied to the knot, it becomes almost impossible to untie. Pieces of wood or pipe inserted through the two loops or wings of the knot will make it easier to untie.

Special knots.

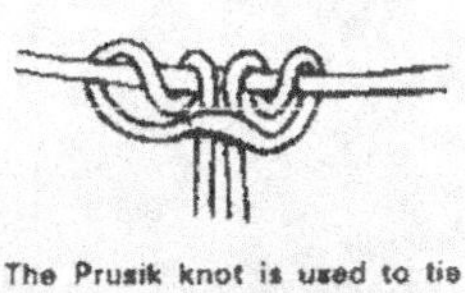

The **Prusik knot** is used to tie one rope around another. One rope will slide along the other if there is no tension, and it will hold if tension is applied. This knot should be tied off with a bowline.

Bowline on a bight forms a double loop.

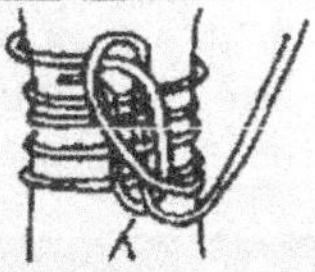

Bowline on a coil can be used by the first and last man on a climbing rope to take up unnecessary slack. A half-hitch must be employed behind the knot.

Section III. MOVING DOWN HILLS AND CLIFFS

RAPPELLING

Rappelling is a means to move quickly down very steep hills and cliffs. Rappelling involves sliding down a rope which has been anchored around a firm object (anchor point) such as a tree, projecting rock, or piton.

ROPE AT ANCHOR POINT

When setting up a rappel site:

Be sure the rope reaches the bottom of the site.

Test the anchor point carefully. The rope should run around the anchor point when pulled from below to recover the rope. Insure that the area is clear of loose rocks.

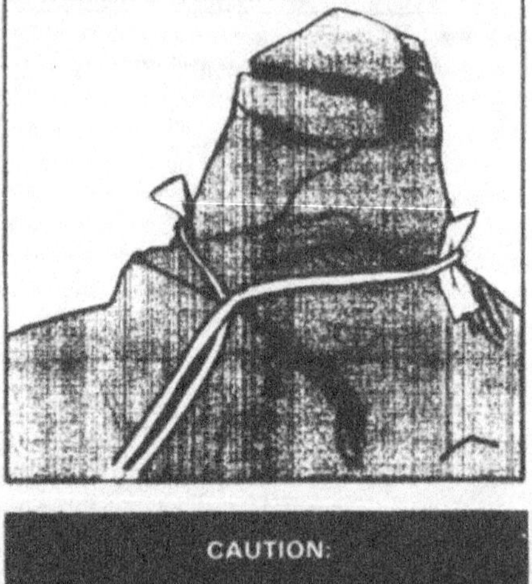

CAUTION:

GLOVES MUST BE WORN DURING ALL RAPPELLING TO PROTECT THE HANDS FROM SEVERE ROPE BURNS.

HASTY RAPPEL

The hasty rappel is a fast, easy way to get down a moderately steep slope or cliff. A soldier must wear a shirt to do a hasty rappel.

HASTY RAPPEL

To conduct a hasty rappel:

Face slightly sideways to the anchor point and place the ropes across the back. The hand nearest the anchor is the guide hand, and the lower hand is the brake hand.

Walk sideways down the hill or cliff, letting the rope move through the hands and across the back.

To stop, bring the brake hand across in front of the body, locking the rope. At the same time, turn and face up toward the anchor point.

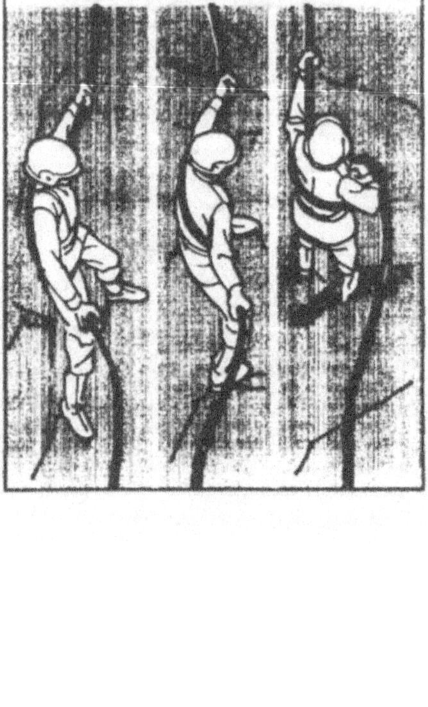

SEAT-HIP RAPPEL

The seat-hip rappel is a fast method of getting down a steep hill or cliff, and it is also used to rappel from helicopters. In this rappel, friction is taken up by a snaplink inserted in a rappel seat fastened to the body.

SEAT-HIP RAPPEL SEQUENCE

Tuck the fatigue jacket into the trousers. Any loose clothing or equipment around the waist can be pulled into the snaplink during the rappel, and present a safety hazard.

Place the center of the sling rope on the hip opposite to the brake hand. (For right-handed soldiers, the right hand is the brake hand. For left-handed soldiers, the left hand is the brake hand.)

Wrap the rope around the waist (keeping the center of the rope on the hip) and tie an overhand knot in front of the body.

Bring the ends of rope between the legs (front to rear), up and around the legs (rear to front), and under the waist loop.

Tie the ends of the rope with a square knot and two half-hitches on the side opposite the brake hand (secure any loose ends).

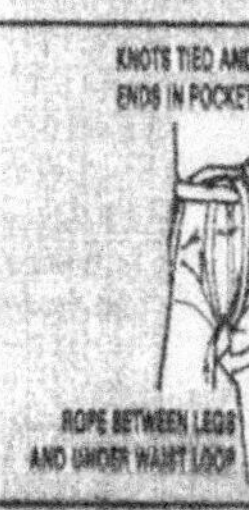

Place the snaplink through the single rope around the waist and through the two ropes forming the overhand knot. (Insert the snaplink with the gate down and the opening toward the body.)

Then rotate the snaplink one-half turn so that the gate is up and opens away from the body.

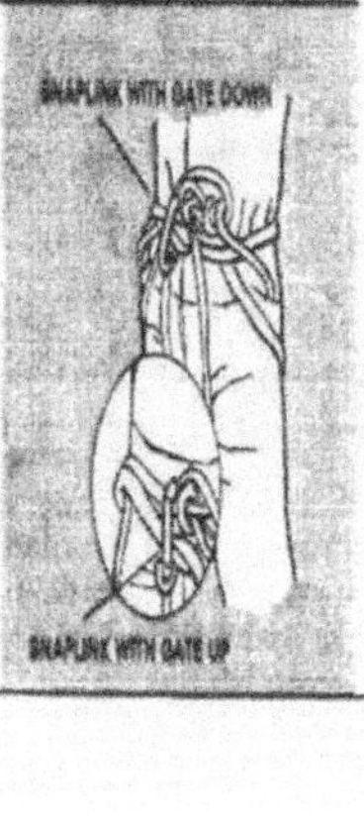

Stand to one side of the rappelling ropes (when braking with the right hand, stand on the left side, and when braking with the left hand, stand on the right). Place the ropes into the snaplink.

Take up some slack in the ropes between the snaplink and the anchor point and bring it underneath, around, over, and back through the snaplink. (This results in a turn of the ropes around the solid shaft of the snaplink which does not cross itself when under tension. When using a single rope, make two turns.)

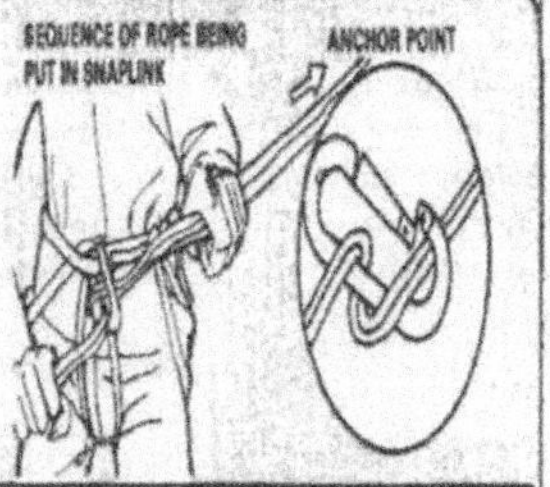

Facing the anchor point, back carefully over the edge of the obstacle and lean well out, almost at a right angle to the surface (the "L" position).

Shout "On rappel."

Go down using the upper hand as a guide and the lower hand as the brake hand. Hold the brake hand behind and slightly above the hip. Brake by closing the brake hand and pressing the rope against the body.

Continue to "walk" down, looking at the ground over the brake hand. To go faster, push off the cliff surface and go down in "bounds" with the brake hand extended toward the ground.

NOTE: For clarity, the load-carrying equipment and weapon are not shown.

Upon reaching the bottom, get off the rope, straighten it, and give the signal, "Off rappel." (When silence is required, use a prearranged signal such as pulling on the rope a set number of times.)

When the last man is down, recover the rope by pulling one end of it until the entire rope has pulled free of the anchor point. This can only be done if the rope is not tied to the anchor point. Pull the rope smoothly to prevent it from getting tangled.

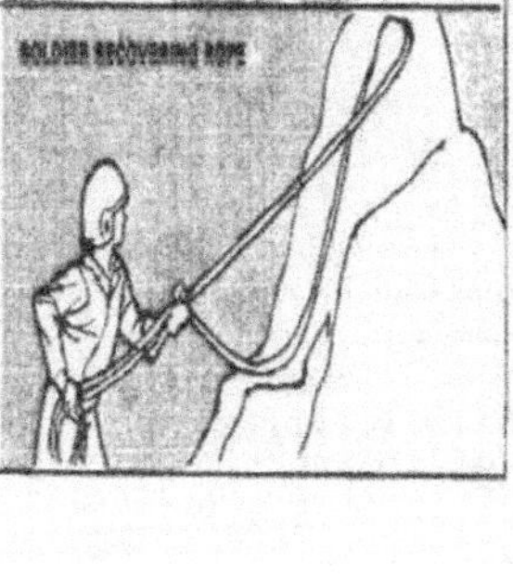

Section IV. MOVING BY HELICOPTER

RAPPELLING FROM HELICOPTERS

In the jungle, there are many places where the vegetation or the ruggedness of the terrain will not permit a helicopter to land. Therefore, it maybe necessary to rappel from a helicopter to get on the ground. Special equipment is required to rappel from helicopters. There are no safe field expedients. The following paragraphs describe the equipment and procedure used in rappelling from helicopters and furnish information on other extraction means.

EQUIPMENT

The *donut ring* is the primary anchoring device (anchor point) inside a helicopter. The floating safety ring is the secondary anchor point. The donut ring consists of a 12-inch solid ring of VS-inch cold-rolled steel cable; seven parachute static line snap hooks; four Ma-inch U-bolts; and 12 inches of chain or '/2-inch cable. Instructions for making a donut ring and a floating safety ring are found in FM 57-38.

The *log coil* helps the double rappelling rope fall clear of the aircraft. To prepare it, start with the running end of the rope and coil the rope evenly and tightly around the log. Use a log approximately 2 to 3 inches in diameter and 16 to 24 inches long.

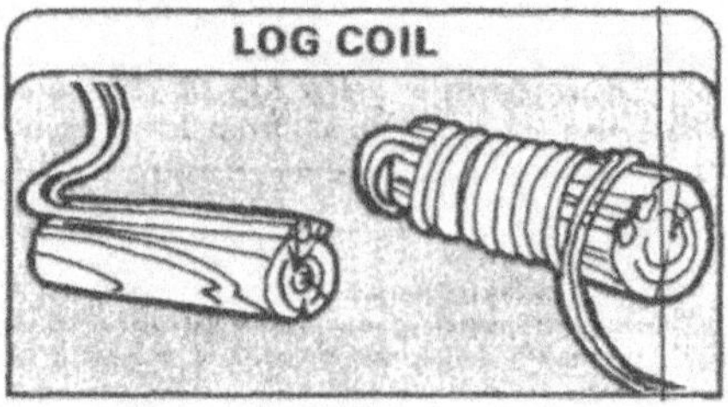

A system that has proven to be almost ideal for rope deployment in helicopter rappel operations is that of the rope deployment bag. The rope deployment bag is issue as a component of the Stabo extraction system The ropes are prepared in the normal rappel configuration and stretched to full length The D-bag is placed at the loose ends of rope opposite the snaplinks. After insur that all rubber bands are present all stowing lines of the D-bag, a bight is formed in the two runnings ends of the rappel ropes and then placed in the center retainer band just above the stow pocket. The rope is then folded and stowed in the retainer bands working from side to side of the D-bag, while making sure that folds do not extend past either side of the D-bag. Six to eight folds of rope are placed in each retainer band, working towards the top of the D-bag. Once the top of the bag is reached, a bight is formed in the climbing ropes 24 inches below the first snaplink and stowed in the top center retainer band. After the D-bag is inspected, the bag is rolled, going from bottom to top, leaving the snaplinks exposed. Tape is used to secure the top flap of the bag.

PROCEDURES

The rappelling rope is connected to the donut ring and the floating safety ring, in the following order, and in the manner described.

CONNECTION OF A RAPPELLING ROPE

The No. 1 snaplink is attached to the donut ring in the following manner:

Take a bight (loop of rope) approximately 5 feet from the end of the working end of the rope.

Insert the rope into the snaplink.

Make one turn through the snaplink, forming a round turn.

Secure the round turn to the snaplink with two half-hitches.

NOTE: Again, for clarity, a single rope is shown. But a double rope would be actually used.

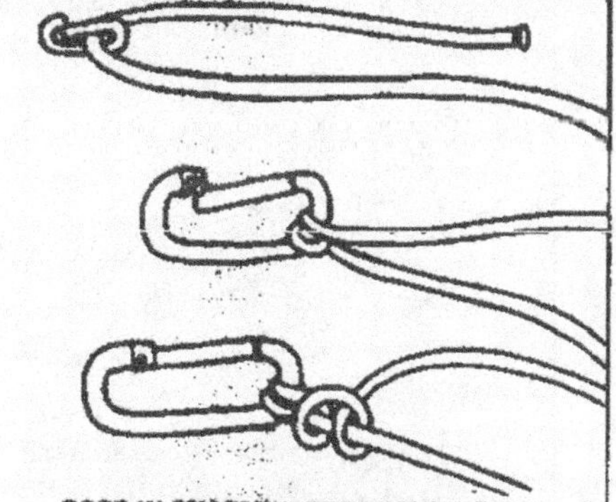

ROPE IN SNAPLINK WITH ROUND TURN AND TWO HALF-HITCHES

Snap the snaplink (gate upward and facing away from the knot) to the donut ring cable.

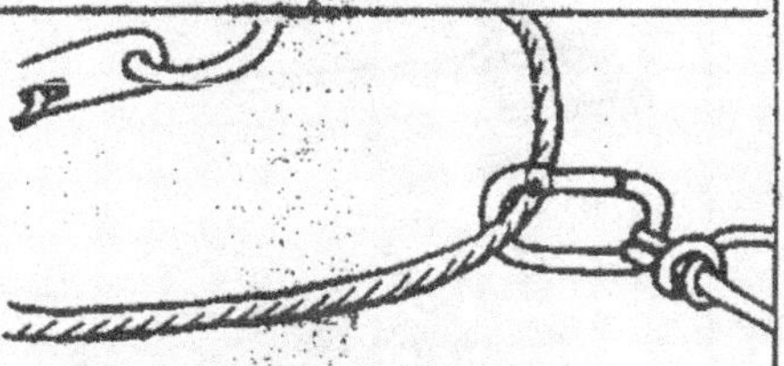

SNAPLINK SNAPPED TO CABLE

The No. 2. snaplink is attached to the floating safety ring in the same manner as the No. 1 snaplink, with the following exceptions:

Take a bight approximately 2 feet from the end of the working end of the rope.

Connect the snaplink to the rope in the same manner as the first connection. Tape the end of the working end of the rope and the knots with masking tape to secure them in place.

Snap the snaplink (gate upward and facing away from the knot) to the floating safety ring.

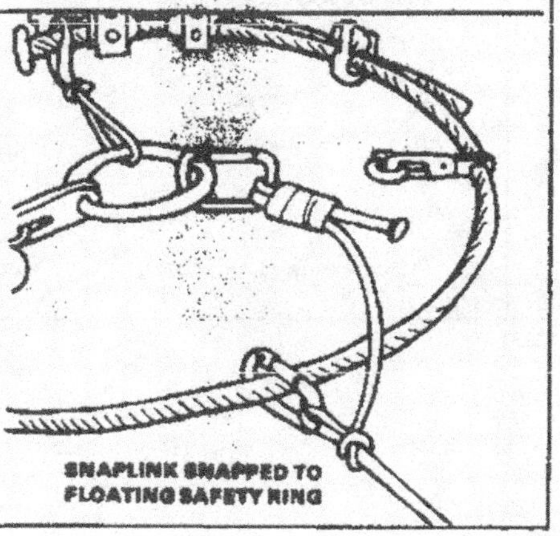

SNAPLINK SNAPPED TO FLOATING SAFETY RING

To rappel from a helicopter,

- Hook into the rope, as in the seat-hip rappel, upon entering the helicopter. Sit on the floor, keeping the brake hand firmly in the small of the back.

- Place the rope in lap.

- Upon the command, "GET READY," look toward the donut ring and pull on the rope to check the anchor point connection. Check the rappel seat and snaplink to insure that the rope is properly inserted. Conduct a final visual inspection of the hookup.

- Upon the command, "SIT IN THE DOOR," swing the feet out to the helicopter skid, keeping the brake on.

- Upon the command, "DROP ROPE," drop the rope with the guide hand, insuring that the rope does not fall between the cargo compartment and the skid and that the rope is not tangled or fouled.

- Upon the command, "POSITION," using the guide hand to assist, pivot 180 degrees on the helicopter and skid bar. Face the inside of the helicopter. Spread the feet shoulder-width apart; lock the knees; and bend forward at the waist, forming an "L" body position.

- Upon the command, *"Go,"* flex the knees and jump backward, letting the rope run through both the brake hand and the guide hand. Descend 5 to 10 meters at a time, looking at the ground over the brake hand. Keep the feet together and legs straight, while maintaining the "L" body position.

- Upon reaching the ground, back all the way out of the rope and move quickly away from beneath the helicopter.

OTHER MEANS OF ENTERING AND LEAVING HELICOPTERS

The *troop ladder* is a good method for larger groups to enter or leave an area where the helicopter cannot land. Installation of the ladder is an aviation responsibility. The crew chief will control the number of troops on the ladder. Only five or six troops will be allowed on the ladder at a time.

Use the legs for climbing and descending; the arms for stability and holding the ladder close to the body.

If possible, each soldier, and especially those carrying heavy loads, should tie a rappel seat with a snaplink attached before ascending. Then, if he becomes tired, he can "snap in," avoiding the chance of a fall.

When going down a ladder, the first soldier on the ground steadies the ladder for the remaining troops. When climbing up a ladder, the soldier designated to hold the ladder steady is the last one up the ladder. If the helicopter starts going up before everyone is loaded, the soldier holding the ladder on the ground should release it at once.

When on the ladder, remain calm at all times. If the helicopter should start settling to the ground, stay calm, watch the ground, and stay on the ladder until reaching the ground.

Once on the ground, move from underneath the helicopter.

The *jungle penetrator* is a metal seat rescue assembly attached to a winch on a helicopter. The seats remain folded when the device is being lowered. The seats are unfolded when the device is on the ground. A conical nose allows it to penetrate the jungle foliage. One to three soldiers (two is a standard load) can be placed on it. Nylon straps are placed under the arms of the soldier(s) to be hoisted. The device lifts the soldier(s) into the helicopter.

TROOP LADDER

JUNGLE PENETRATOR

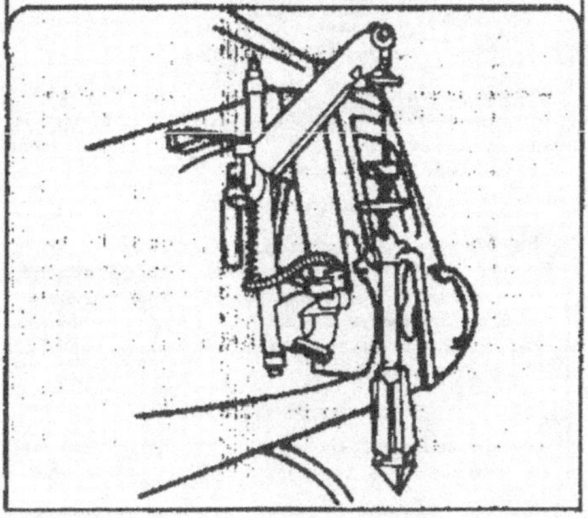

WARNING:

WHILE THE PENETRATOR IS BEING LOWERED. STATIC ELECTRICITY CAN BUILD UP ON THE METAL SURFACE. THE PENETRATOR SHOULD TOUCH THE GROUND BEFORE ANYONE TOUCHES IT, TO ALLOW THE STATIC ELECTRICITY TO GROUND ITSELF

OTHER EXTRACTION MEANS

Other means which can be used for extraction of troops where landing zones (LZ) are not available include such devices as the personnel Stabo extraction system, the Maguire rig, the Palmer rig, or the jungle operations extraction system (JOES).

STABO SYSTEM

This system provides a means for rapid pickup of soldiers by helicopter from areas where the helicopter cannot land. The system consists of the *personnel harness,* the *bridle, the suspension rope,* the *safety rope,* and the *deployment bag.*

As the pickup process is initiated, the helicopter hovers over the pickup zone at

STABO SYSTEM

PERSONNEL HARNESS

The **personnel harness** is designed to be worn by the user in the field and to partially replace the soldier's load-carrying equipment. It is made of nylon webbing and stitched to a standard, medium, or large web pistol belt. The two V-rings at the top of each of the harness shoulder straps connect to the bridle snaphooks and provide a lift point for the harness. The leg straps are adjustable for comfort. When the user is in the field, leg straps may be disconnected, folded, and secured to a suitable point on the harness.

BRIDLE

The **bridle** is also made of nylon webbing. During use, the D-ring is connected to a suspension rope snaphook while the two bridle snaphooks are attached to the two personnel harness lift V-rings.

SUSPENSION ROPE

The **suspension rope** is the same nylon rope that is used for rappelling. It has snaphooks attached to both ends: one to connect to the bridle and one to connect to the anchor point in the helicopter.

SAFETY ROPE

The **safety rope** is 3 meters long and made of nylon. Each end is looped and spliced, with a snaphook attached to each loop. It is used when two or three soldiers are extracted together, and it provides a means for them to hold together to minimize wind buffeting and oscilation during pickup and flight.

DEPLOYMENT BAG

The **deployment bag** is made of cotton duck and contains the bridle, suspension rope, and safety rope.

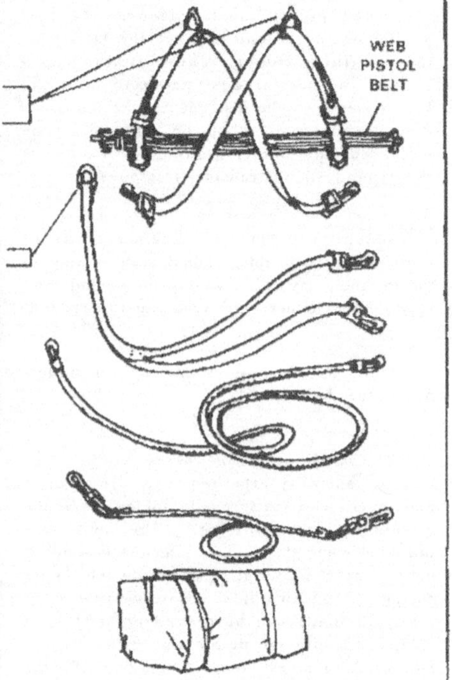

altitudes up to 150 feet. A member of the helicopter crew drops the extraction system deployment bag from the left door of the helicopter. (A maximum of three units may be connected and dropped simultaneously.) As the deployment bag descends, the suspension rope deploys until the bag reaches the ground. The soldier to be extracted then attaches the bridle snaphooks to the lift V-rings on his harness. After insuring that the leg straps are connected and tight the soldier notifies the helicopter by radio or hand signals that liftoff may begin. The helicopter then lifts the soldier from the area and, carrying him suspended beneath the helicopter, moves to an area where a safe landing can be made. The helicopter then lowers the suspended soldier to the ground, lands nearby, and allows him to board the helicopter. For details on employing this system, consult *TM 10-1670-262-12.*

SOLDIER IN HARNESS BEING LIFTED OFF

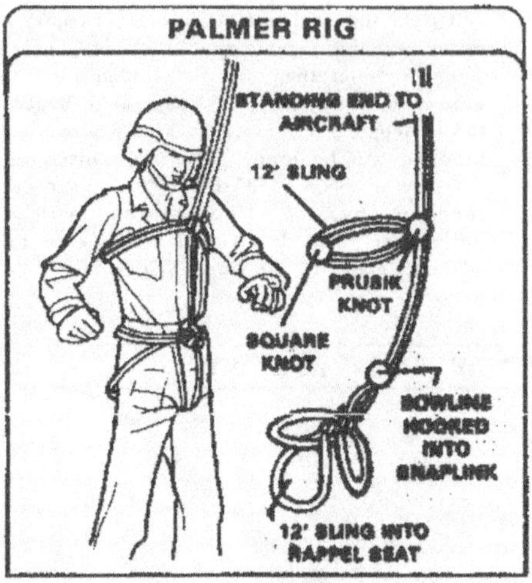

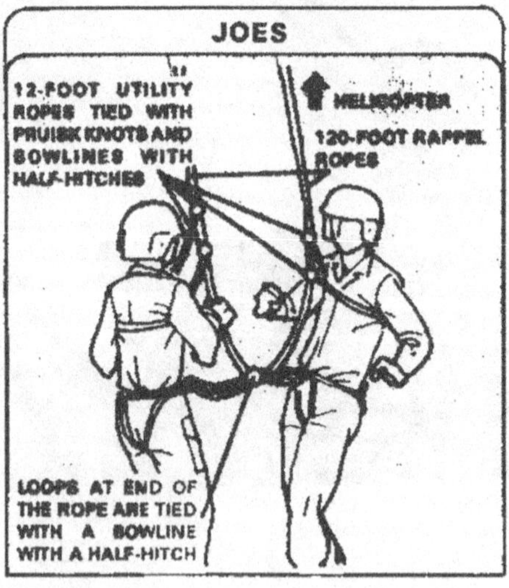

PALMER RIG

This rig is constructed with the 120-foot nylon rope and two 12-foot nylon sling ropes. One end of the 120-foot rope is secured to the donut ring. The running end of the 120-foot rope is tied with an end of the rope bowline knot at the end of the rope. One of the 12-foot sling ropes is tied 3 feet above the bowline with a Prusik knot, and the loose ends are tied off with a square knot to form a loop. The rider uses the remaining sling rope to form a rappel seat. He then fastens a snaplink to the rappel seat and the bowline knot; places his arms up and through the upper loop; and is extracted. The Palmer rig is relatively safe and may be used when evacuating wounded personnel.

MAGUIRE RIG

This rig is also simple and easy to construct. It is made with an 8-foot by 2-inch piece of nylon webbing sewn together at the ends to form a loop containing a D-ring. A smaller slip loop (wrist loop) is sewn 12 inches down from the top of the larger loop. During extraction, the rider simply sits in the seat of the large loop. He then places his wrist in the slip loop and tightens the loop, insuring that he does not fall from the larger loop during extraction.

JUNGLE OPERATIONS EXTRACTION SYSTEM

This system was developed after careful evaluation of the three previously mentioned systems, while considering those items of equipment available to the individual soldier. The JOES can be quickly constructed from components readily available from the supply system. The individual soldier requires very little (10 to 15 minutes) training in what he will be required to do if he needs to be extracted, and only a maximum of 5 minutes is required for him to actually prepare for extraction.

CONSTRUCTION OF A JOES FOR TWO INDIVIDUALS

Material needed:

- Anchor system for helicopter (donut ring).
- Two 120-foot rappel ropes.
- Three military snaplinks.
- Two 12-foot utility ropes.
- One deployment system.

Once all equipment has been thoroughly inspected, the JOES is prepared in the following manner.

- Lay out both 120-foot rappel ropes side by side.
- Move to one end and, taking both ropes as one, tie a round turn and two half-hitches around a snaplink approximately 5 feet from the end.
- From the first knot, move 18 inches toward the short end of the rope and tie another round turn and two half-hitches around another snaplink. Tape any excess rope. This now becomes the fixed end of the rope and will be attached to the anchor system in the helicopter.
- Move to the working end of the rope and, taking both ropes as one, tie a bowline with a half-hitch. The loops formed by this bowline should be approximately 12 inches in diameter. Any excess over 6 inches should be taped to help eliminate confusion during hook up.
- Approximately 18 inches from the top of the bowline knot toward the fixed end of the rope, take one end of a 12-foot utility rope and tie a Prusik knot and a bowline with a half-hitch around one 1 20-foot rappel rope.
- On the same rope, using the other end of the 12-foot utility rope, tie the same knot directly below the first.
- Repeat steps 5 and 6 with another sling rope on the other 120-foot rappel rope.
- Pack JOES in D-bag.

Each individual who is to be extracted must have a 12-foot utility rope and a snaplink. If the individual does not have these items, they can be dropped to him with the JOES. The individual to be extracted makes a rappel seat out of the 12-foot utility rope and installs the snaplink as if he were going to make a rappel.

When JOES is dropped, he and his buddy move to the bag. Each individual hooks his snaplink into one loop at the end of the rope bowline, and places the loop formed by one of the 12-foot utility ropes over his shoulders and under his armpits. He and his buddy stay as far back from the helicopter as practical until it starts to lift up, and as tension is put on the rope, they move forward until they are directly underneath it and linked up with other personnel. All soldiers should link up by holding onto the adjacent person's equipment.

Section V. MOVING ACROSS WATER OBSTACLES

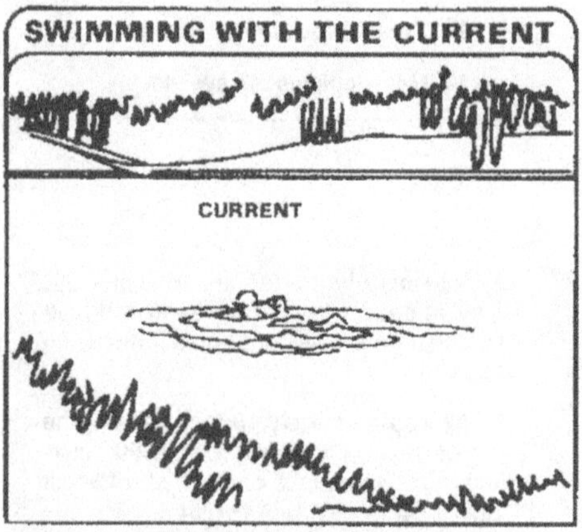

CROSSING RIVERS AND STREAMS

There are several expedient ways to cross rivers and streams. The ways used in any situation depends on the width and depth of the water, the speed of the current, the time and equipment available, and the friendly and enemy situation.

There is always a possibility of equipment failure. For this reason, every soldier should be able to swim. In all water crossings several strong swimmers should be stationed either at the water's edge or, if possible, in midstream to help anyone who gets into trouble.

If a soldier accidentally falls into the water, he should swim with the current to the nearer bank. Swimming against the current is dangerous because the swimmer is quickly exhausted by the force of the current.

FORDING

A good site to ford a stream has these characteristics:

- **Good concealment on both banks.**
- **Few large rocks in the river bed. (Submerged large rocks are usually slippery and make it difficult to maintain footing.)**
- **Shallow water or a sandbar in the middle of the stream. Troops may rest or regain their footing on these sandbars.**
- **Low banks to make entry and exit easier. High banks normally mean deep water. Deep water near the far shore is especially dangerous as the soldiers may be tired and less able to get out.**

A unit should cross at an angle against the current. Each soldier should keep his feet wide apart and drag his legs through the water, not lift them, so that the current will not throw him off balance. Poles can be used to probe in front of the troops to help find deep holes and maintain footing.

FLOATING AIDS

For deeper streams which have little current, soldiers can use a number of floating aids such as the following:

- The standard air mattress
- Trousers

NOTE: Trousers must be soaked in water before using.

- Canteen safety belt
- Poncho life belt
- Water wings
- Poncho brush raft
- Australian poncho raft
- Log rafts

TROUSERS AS A FLOATING AID

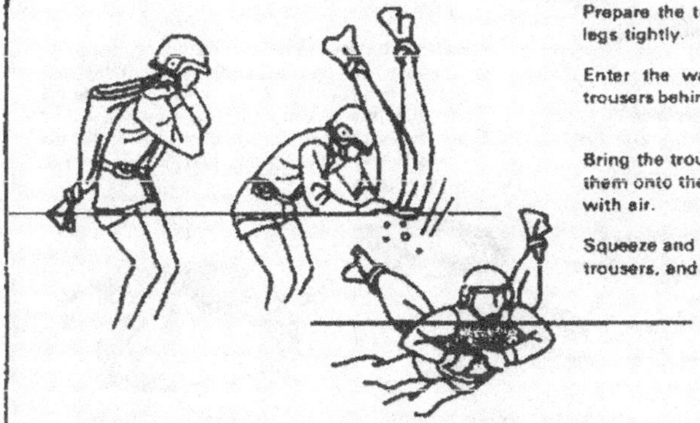

Prepare the trousers by tying the bottoms of the legs tightly.

Enter the water to waist depth and hold the trousers behind the shoulders, with the waist open.

Bring the trousers quickly over the head and bang them onto the water in front. This action fills them with air.

Squeeze and hold the waist together, lie over the trousers, and float as if on water wings.

CANTEEN SAFETY BELT

Attach at least eight empty plastic canteens to a pistol belt (or tie them to a rope which can then be used as a belt). Insure that the caps are screwed on tightly.

PONCHO LIFE BELT

Roll green vegetation tightly inside a poncho and fold the ends over to make a watertight life belt. Roll up the life belt like a big sausage at least 8 inches in diameter and tie it. Wear it around the waist or across one shoulder and under the opposite arm like a bandoleer.

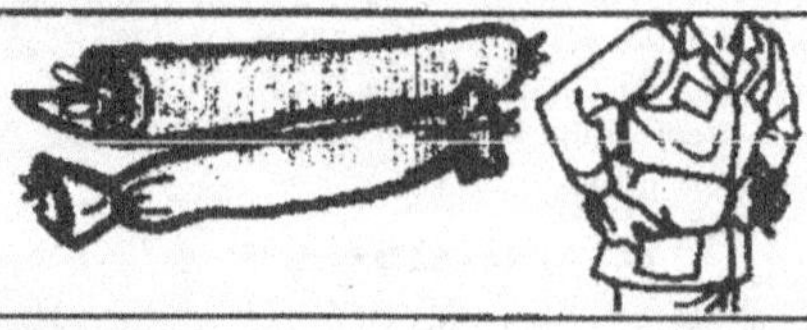

WATER WINGS

Two or more air-filled plastic bags, securely tied at the mouth, can be used as expedient water wings. Other expedients include empty water or fuel cans and ammunition canisters.

CONSTRUCTION OF PONCHO BRUSH RAFT

Use two ponchos, and tie the neck of each tightly by using the drawstring.

Spread one poncho on the ground with the hood up so that it will end up inside the raft.

Cut fresh, green brush (avoid thick branches or wood stakes) and pile it on the poncho to a height of 18 inches.

Place an X-frame made of small saplings (1 to 1 1/2 inches in diameter and 3 to 4 feet long) on the brush. Anchor this frame by tying the drawstring of the poncho to the center of the X-frame.

Pile another 18 inches of brush on top of the X-frame.

Compress the brush slightly and fold up the poncho, tying ropes or vines diagonally across the corner grommets and straight across from side grommets. The sides of the poncho should not touch.

Spread the second poncho on the ground, with hood up, next to the bundle made of the first poncho and brush. Roll the bundle over onto the center of the second poncho and tie the second poncho across the sides and diagonally across the corners. This raft will safely float 250 pounds and is very stable.

CONSTRUCTION OF AUSTRALIAN PONCHO RAFT

When there is not enough time to gather a lot of brush, this raft is made by using a soldier's combat equipment for bulk. Normally, two soldiers make this poncho together. It is more waterproof than the poncho brush raft but will float only about 80 pounds of weight. Two soldiers make this raft as follows.

Place one poncho on the ground with the hood facing up. Close the neck opening by tying it off with the drawstring.

Place two poles (or branches), about 1 to 1 1/2 inches in diameter and 4 feet long, in the center of the poncho about 18 inches apart.

Next, place the rucksack, and any other equipment desired, between the poles.

EQUIPMENT ON PONCHO

Snap the poncho together. Hold the snapped portion of the poncho in the air and roll it tightly down toward the equipment. Roll from the center out to both ends. At the ends, twist the poncho to form "pigtails." Fold the pigtails inward toward each other and tie them tightly together with boot laces, vines, communication wire, or other available tying material.

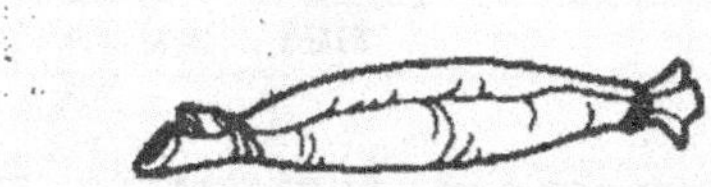

PONCHO ROLLED WITH PIGTAILS

Spread the second poncho on the ground, neck closed and facing up.

Place the equipment bundle formed with the first poncho, with the seam (tied pigtails) facing down, on the second poncho.

Roll and tie the second poncho in the same way as the first.

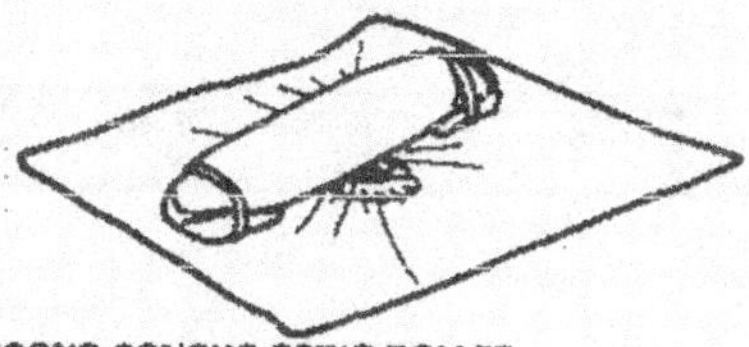

SECOND PONCHO BEING ROLLED

An empty canteen tied to one end of a rope with the other end tied to the raft helps in towing. One soldier pulls on the rope while the other pushes the raft. Place weapons on top of the raft and secure them with ropes. The weapons should be secured to the raft by the use of quick releases. The raft is now ready for the water.

SOLDIERS MOVING RAFT IN WATER

When launching any poncho raft or leaving the water with it, take care not to drag it on the ground as this will cause punctures or tears.

CONSTRUCTION OF LOG RAFT

Logs, either singly or lashed together, can be used to float soldiers and equipment. Be careful when selecting logs for rafts. Some jungle trees will not float. To see whether certain wood is suitable, put a wood chip from a tree in the water. If the chip sinks, so will a raft made of that wood.

ROPE BRIDGES

For crossing streams and small rivers quickly, rope bridges offer a suitable temporary system, especially when there is a strong current. Because of the stretch factor of nylon ropes, they should not be used to cross gaps of more than 20 meters. For larger gaps, manila rope should be used.

DETERMINATION OF RIVER WIDTH

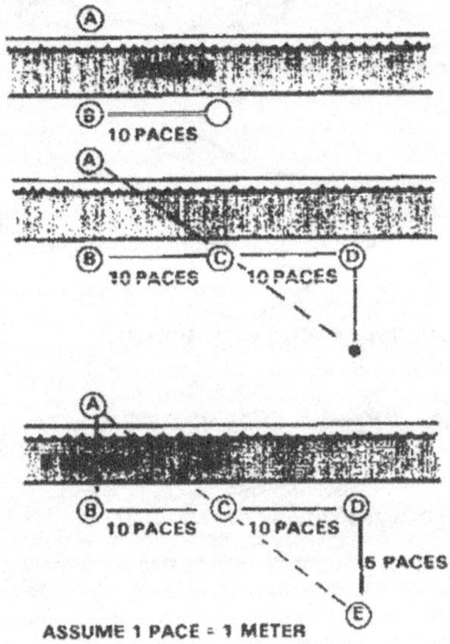

A method used to measure the width of a river or stream is described below.

Select a straight section of the stream.

Pick two points opposite each other (A and B).

Stand at B; turn in a direction parallel to the stream; walk off 10 paces. Mark that point as point C (B to C = 10 paces).

Continue walking in the same direction 10 more paces. Mark that point as point D (C to D = 10 paces).

Turn at a right angle away from the stream and walk until you are on line with points C and A. Mark this point as point E. Determine the distance between D and E by converting the pace count into meters. In this example, 1 pace is equal to 1 meter and the pace count is 5 paces. Therefore, the distance between D and E is 5 meters.

The distance from D to E is equal to the distance from A to B. Therefore, the width of the stream is also about 5 meters.

In order to erect a rope bridge, the first thing to be done is to get one end of the rope across the stream. This task can be frustrating when there is a strong current. To get the rope across, anchor one end of a rope that is at least double the width of the stream at point A. Take the other end of the line upstream as far as it will go. Then, tie a sling rope around the waist of a strong swimmer and, using a snaplink, attach the line to him. He should swim diagonally downstream to the far bank, pulling the rope across.

SWIMMER PULLING ROPE ACROSS

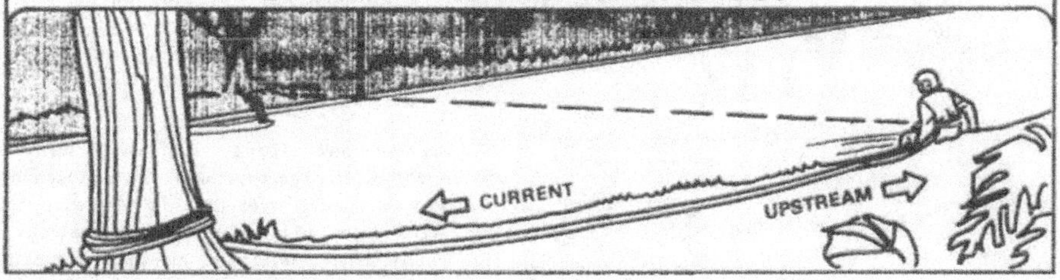

One-Rope Bridge. A one-rope bridge can be constructed either above water level or at water level. The leader must decide which to construct. The bridge is constructed the same regardless of the level.

CONSTRUCTION OF ONE-ROPE BRIDGE

To construct this bridge:

The bridge rope is pulled around the upstream side of the far side anchor point and temporarily secured without tying a knot. On the near side, a transport-tightening system is placed in the bridge rope by tying a double butterfly knot and placing two snaplinks in the butterfly. The running end of the bridge rope is then passed around the downstream side of the near side anchor point and through the two snaplinks.

When this transport-tightening system is prepared, the soldier on the far side pulls the butterfly knot approximately a third of the distance across the river. He then secures the bridge rope to the far side anchor point using a round turn and two half-hitches. Soldiers on the near side then pull the slack out of the bridge rope until the butterfly knot is back on the near side. (The bridge rope must be as tight as possible so that it will not sag when used.) The bridge rope is then tied off against itself using two half-hitches with a quick release in the last half-hitch.

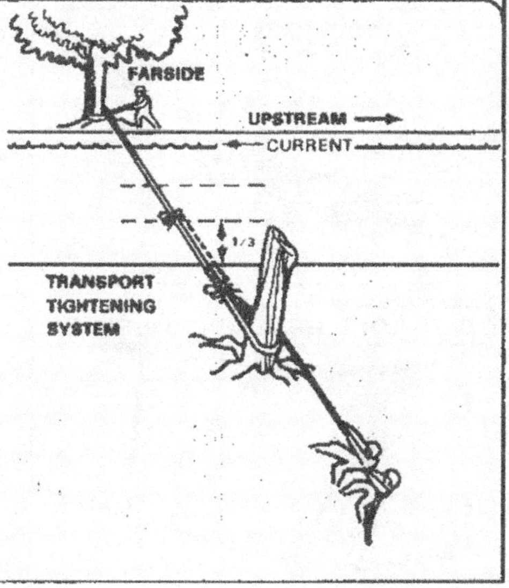

Crossing Method above Water Level. Use one of the following methods.

Commando crawl. Lie on the top of the rope with the instep of the right foot hooked on the rope. Let the left leg hang to maintain balance. Pull across with the hands and arms, at the same time pushing on the rope with the right foot. (For safety, each soldier ties a rappel seat and hooks the snaplink to the rope bridge.)

Monkey crawl. Hang suspended below the rope, holding the rope with the hands and crossing the knees over the top of the rope. Pull with the hands and push with the legs. (For safety, each soldier ties a rappel seat and hooks the snaplink to the rope bridge.) This is the safest and the best way to cross the one-rope bridge.

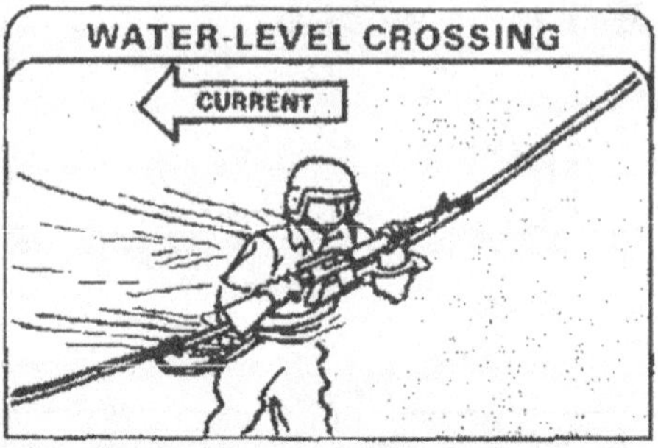

Crossing Method at Water Level. Hold onto the rope with both hands, face upstream, and walk into the water. Cross the bridge by sliding and pulling the hands along the rope. (For safety, each soldier ties a sling rope around his waist, leaving a working end of about 3 to 4 feet. He ties a bowline in the working end and attaches a snaplink to the loop. He then hooks the snaplink to the rope bridge.)

To recover the rope, the last soldier unties the rope, ties it around his waist and, after all slack is taken up, is pulled across.

Two-rope bridge. Construction of this bridge is similar to that of the one-rope bridge, except two ropes, a hand rope and a foot rope, are used. These ropes are spaced about 1.5 meters apart vertically at the anchor points. (For added safety, make snaplink attachments to the hand and foot ropes from a rope tied around the waist. Move across the bridge using the snaplink to allow the safety rope to slide.) To keep the ropes a uniform distance apart as men cross, spreader ropes should be tied between the two ropes every 15 feet. A sling rope is used and tied to each bridge rope with a round turn and two half-hitches.

OTHER CROSSING MEANS

Suspension traverses, bridges, and cableways can be used to move large numbers of soldiers or heavy equipment over wide rivers and ravines, or up and down cliffs in a short period of time. Because heavy or bulky material or equipment is needed to construct these expedients, their use is practical only if the needed items can be transported to the site by air or surface means (watercraft, pack animals, etc.).

APPENDIX D

Waterborne Operations

Section I. GENERAL

The inland waterways and jungle coastal or delta regions are land environments dominated by water routes. There may be one or more major waterways and an extensive network of smaller waterways. Usable roads are scarce, and cross-country movement is extremely difficult. The following describes jungle waterways.

UPPER SECTOR (HEADWATERS)

The headwaters of a waterway are usually formed in a mountainous region. The headwaters consist of numerous tributaries which merge to form a river system as the water flows down to the valley. Headwaters are characterized by waterfalls, rapids, and variations in water depth, all of which restrict the use of watercraft.

MIDDLE SECTOR (CENTRAL VALLEY)

When the waterway reaches the central valley, it has formed a broad river which is usually navigable for great distances inland. This river is usually fed by numerous tributaries. In those jungles where there are definite dry and rainy seasons, many of the tributaries found during the rainy season may not exist during the dry season. The river in the valley is wide, slow, and often meanders. During periods of heavy rainfall, the course of the river may change. The jungle vegetation grows up along the riverbanks to form an almost solid wall. The banks of the river are often steep and slippery. Many of the navigable tributaries feeding the major river will often be completely overgrown with vegetation and contain obstacles such as fallen trees.

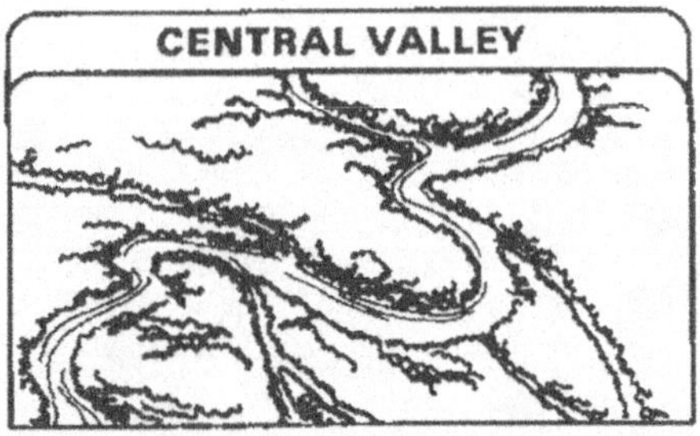

CONTENTS

LOW SECTOR (DELTA)

When the river reaches the low coastal area, it spreads over a flat, alluvial plain and becomes a number of river tributaries (small streams or channels spreading fanlike from the main channel) disbursing a great amount of sediment into a gulf, bay, or ocean. Usually, there are many large and small tidal streams and channels, whose current may change speed or reverse with the tide in a predictable manner. Bottoms of the tributaries normally slope up to a crest or bar at the river's mouth. In some instances, only watercraft with a draft (that part of the craft under water) of 1 to 2 meters will be able to cross the crest or bar at high tide.

section II. PREPARING FOR OPERATIONS

WATERBORNE COMBAT OPERATIONS

The fundamentals and tactics applicable in conventional ground operations apply in waterborne operations. However, special organization, equipment, and techniques are required when ground forces are supported by Navy ships and craft. The waterborne force should be employed with all available modes of transportation to seek out and destroy the enemy and his installations. One portion of the force may enter the area by watercraft; another may enter by helicopters; still another may enter the area by moving overland. All units then maneuver to attack the enemy. All available fire support should be used in the operation: close air support, attack helicopters, waterborne and land-based artillery, and naval gunfire. Special considerations in the conduct of jungle waterborne operations include the following:

- The heavy vegetation along the banks of inland waterways offers excellent concealment and enhances the effectiveness of ambushes against watercraft. Therefore, counterambush measures must be planned in conjunction with all water movements. Steep, slippery river banks coupled with dense vegetation often make committing the waterborne force in a coordinated assault landing extremely difficult.

- Security measures during the movement phase along a jungle waterway include proper watercraft formations, constant water patrolling, and air observation, when possible. Fire support to include mortar, artillery, close air support, and available naval gunfire must be preplanned for all water movements.

- Intelligence is critical along jungle waterways. While aerial reconnaissance yields a considerable amount of information, it will have to be supplemented by reconnaissance by boat, especially in areas where tributaries are overgrown by vegetation.

Section III. USING JUNGLE WATERWAYS

SMALL BOAT HANDLING

The use of inland and coastal waterways can add flexibility, surprise, and speed to tactical operations in jungle areas. Use of these waterways will also increase the load-carrying capacity of units which normally operate dismounted. Thus, every combat leader should be familiar with the tactical and technical aspects of small boat handling.

EQUIPMENT

There are several types of small craft available.

RB-3

Reconnaissance Boat, Inflatable: RB-3

- Length: 3 meters
- Width: 1.3 meters
- Weight: 24 pounds, 33 pounds w/paddles and pump
- Load: A two-man crew and 500 pounds or a three-man crew and 300 pounds.
- Powered by: Paddles.

RB-15

Assault Boat, Pneumatic: RB-15

- Length: 5.2 meters
- Width: 1.8 meters
- Weight: 260 pounds
- Crew: 1 coxswain, 10 paddlers
- Load: A maximum of 15 men (11-man crew and 4 passengers) with equipment or 3,300 pounds.
- Powered by: Paddles or a 25-horsepower outboard motor.

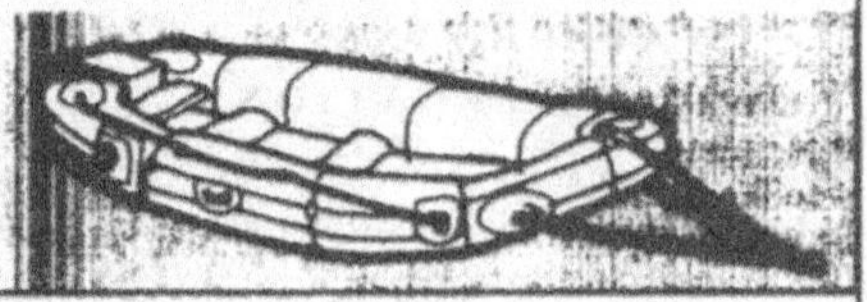

RB-7

Landing Boat Pneumatic: RB-7

- Length: 4 meters
- Width: 2 meters
- Weight: 150 pounds
- Crew: 1 coxswain, 6 paddlers
- Load: 10 men (7-man crew and 3 passengers) with equipment.
- Powered by: Paddles.

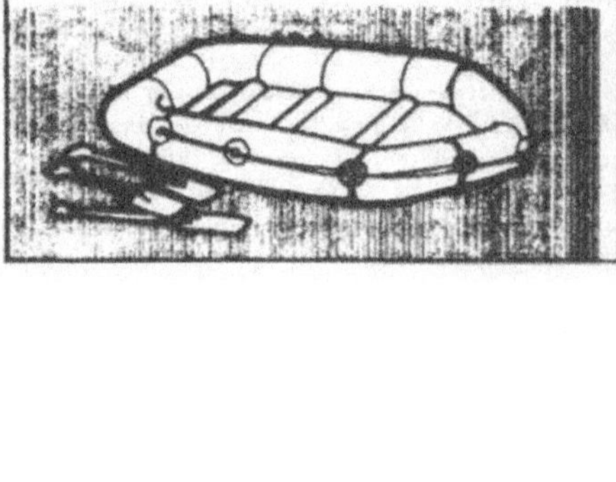

PAB

Plastic Assault Boat, M3: PAB

- Length: 4.9 meters
- Width: 1.7 meters
- Weight: 300 pounds
- Crew: 1 coxswain, 10 paddlers
- Load: A maximum of 15 men (11-man crew and 4 passengers) with equipment or 3,200 pounds.
- Powered by: Paddles or a 25-horsepower outboard motor.

Boats may be powered by outboard motors, paddles, or oars. The mission, availability, and the river itself dictate the method of propulsion to be used.

Motors are noisy. On the other hand, they provide speed, reduce fatigue, and free personnel for security missions. The noise form motors can be heard for distances of 500 to 1,000 meters by day and up to 5,000 meters at night (sound carries better in the quieter, cooler night air). Provided the craft is not seen and the motor does not change pitch, however, it is difficult to estimate the direction of the sound and exactly how far away it is. Bearing these factors in mind, troops may be able to disguise a movement by deceptive tactics such as having other craft work the area. Stopping the motor when traveling downstream will also aid the security of movement.

Paddling is a slow and tiring process, but it is quieter than using motors. (With plastic, wooden, or metal craft, paddles may need to be wrapped with cloth to reduce noise when the paddles strike the craft.)

ORGANIZATION

Before a waterborne operation, each person in a boat is assigned a specific boat position and a corresponding number. (This is the long count method of organization.)

NOTE: The unit, normally a squad, that uses the RB-15 for transportation actually comprises the crew that operates it. All others that do not operate the boat are passengers.

First, the crewmembers are assigned their positions. Next, the passengers are assigned their positions. When using an RB-15, for example, the crewmembers are assigned positions 1 through 11, and the passengers are assigned positions 12 through 15. One person is designated as the boat commander (normally the coxswain). Two persons are designated as a navigator-observer team.

For operational purposes, the crew is organized into pairs. Passengers are not numbered in this method. (This is the short count method of organization.) When using the RB-15, for example, the crewmembers (in pairs) are assigned to operational positions 1 through 5.

PADDLES WRAPPED TO REDUCE NOISE

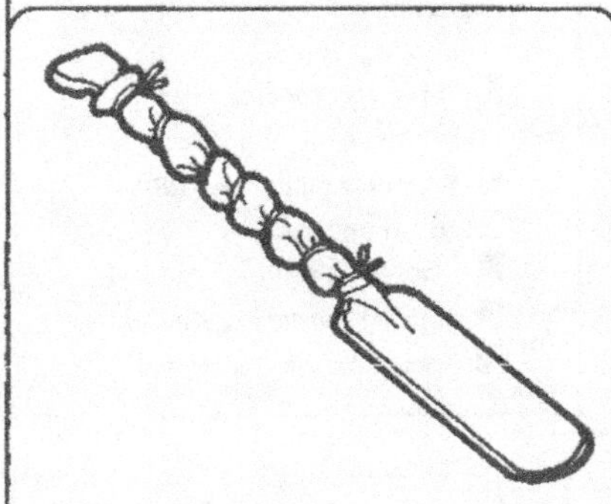

RB-15 WITH POSITIONS NUMBERED—LONG COUNT

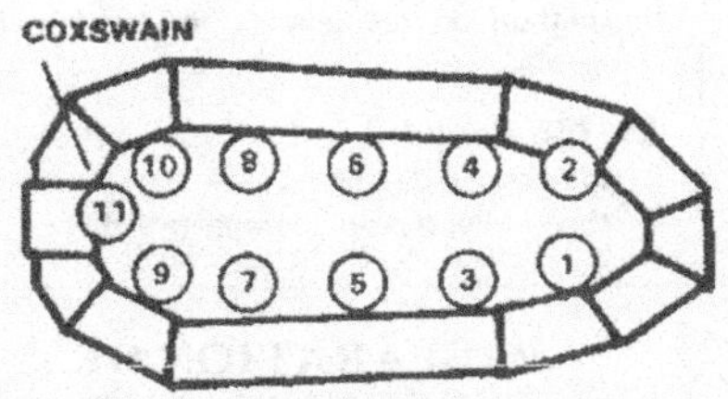

NOTE: PASSENGERS ARE NUMBERED FROM FRONT TO REAR AFTER CREW IS NUMBERED.

RB-15 WITH POSITIONS NUMBERED—SHORT COUNT

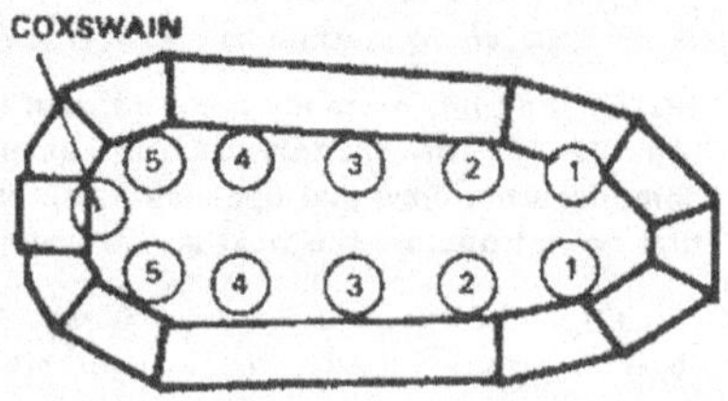

Crew duties:

- The coxswain is responsible for the control of the boat and action of the crew. He supervises the loading, lashing, and distribution of equipment, He also maintains the course and speed of the boat.
- The number 1 paddler (long count method) is the observer and is responsible for the storage and use of the bowline.
- The number 2 paddler (long count method) is responsible for setting the stroke.
- All paddlers are responsible for loading and lashing the equipment in their respective compartment.

PREPARATION OF PERSONNEL AND EQUIPMENT

Each crewmember and passenger must wear a life preserver.

The load-carrying equipment harness is worn unbuckled at the waist.

The rifle is slung outside of thelife preserver, opposite the outboard side, with the muzzle down.

Crew-served weapons, radios, ammunition, and other bulk equipment are lashed securely to the boat to prevent loss if the boat should overturn.

Radios, batteries, and unboxed ammunition are waterproofed.

Hot weapons are cooled prior to being placed in the boat to prevent damage to the boat or injury to personnel.

Pointed objects are padded to prevent puncture of the boat.

The most effective equipment-lashing system that has been developed is the RB-15 lashing system. This system is quick and easy to install, requires no special equipment, and prevents loss of equipment in the event the craft is capsized. It also allows the craft to be easily righted.

The equipment needed in this lashing system is:

Ten sling ropes-

- Three for capsize lines.

One for securing the M60.

- Six for rigging the RB-15.

Nine snaplinks-

- Two for securing the M60.
- Four for securing rucksacks.
- Three for the center line.

The average squad can fully rig and lash an RB- 15, using this system, in approximately 15 minutes.

RB-15 LASHING SYSTEM

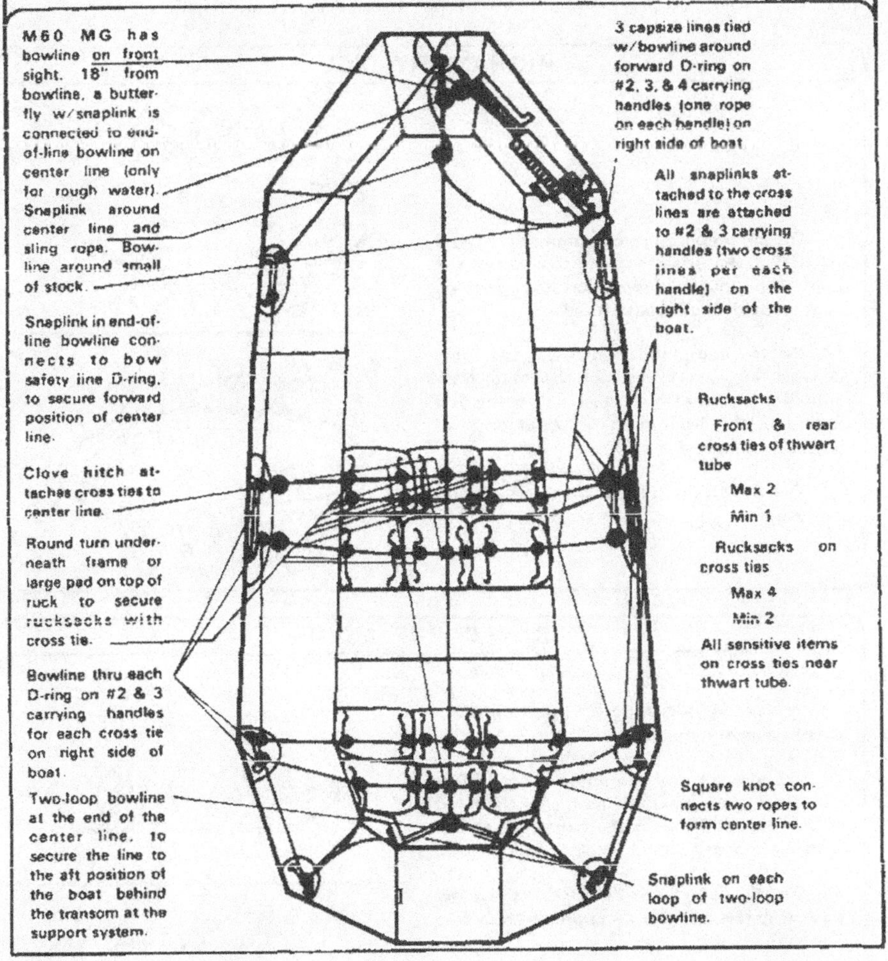

COMMANDS

"short Count, *count off.* " Crew counts *off* their positions by pairs, for example, 1,2,3,4, 5, coxswain (RB-15).

"L°"S *count, count off.* " Crew counts off their positions by individuals, for example, 1, 2, 3, 4, 5> 6, 7, 8, 9, 10, coxswain (RB-15).

"Boat stations." Crew takes position along side of boat.

"High carry, mole" (used for long distance moves overland).

HIGH CARRY, MOVE

On the preparatory command of "**HIGH CARRY,**" the crew faces to the rear of the boat and squats down grasping carrying handles with the inboard hand.

On the command, "**MOVE,**" the crew swivels around, lifting the boat to their shoulders, so that the crew is standing and facing to the front with the boat on their inboard shoulders.

The coxswain guides the crew during movement.

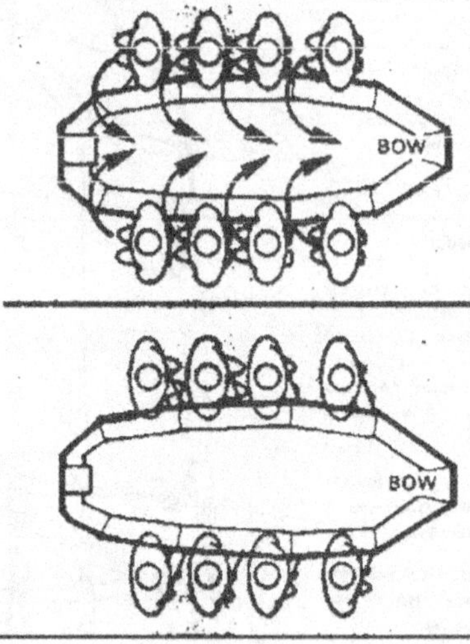

LOW CARRY, MOVE

"**LOW CARRY, MOVE**" (used for short distance moves overland).

On preparatory command of "**LOW CARRY,**" the crew remains facing the front of the boat and grasps the carrying handles with the inboard hand.

On the command, "**MOVE,**" the crew stands up raising the boat approximately 6 to 8 inches off the ground.

The coxswain guides the crew during movement.

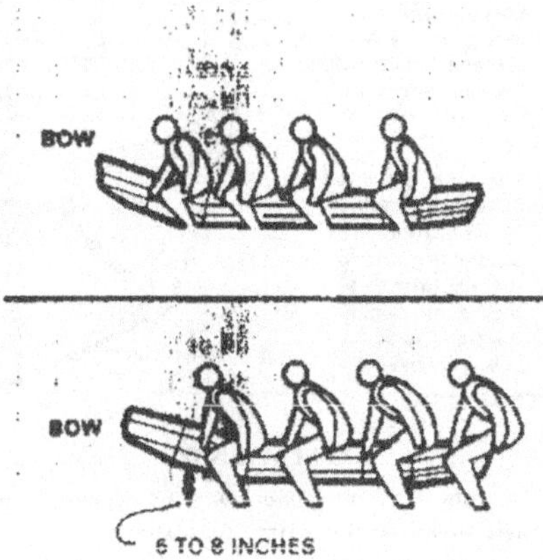

"Lower the boat, move." Crew lowers the boat gently to the ground using carrying handles.

"Give way together." Crew paddles to the front, with Number 2 setting the stroke for the rest of the crew.

"Hold." Entire crew keeps paddles motionless in the water, thereby stopping the boat.

"Hold left (right)." Left crewmembers hold, right crewmembers continue with previous command.

"Back paddle." Entire crew paddles backward. This action propels the boat to the rear.

"Back paddle left (right)." Left crewmembers back paddle causing the boat to turn left, right crewmembers continue with previous command.

"Rest paddles." Crewmembers place paddles on their laps with blades outboard. This command may be given to pairs, i.e., *"Number l's rest paddles."*

LAUNCHING AND LANDING

When launching, the crew maintains a firm grip on the boat until they are inside the boat; similarly, when landing, they hold onto the boat until it is completely out of the water.

The crew stays as low as possible when entering and leaving the boat to avoid capsizing it.

Crewmembers can load or unload a boat either by individuals or in pairs. They load and unload a boat by individuals at steep river banks and along shoreline where the water is deep near the shore. They also load or unload from or into a larger vessel such as a landing craft, mechanized (LCM), by individuals.

The crewmembers load or unload individually according to their number in the long count method. The coxswain directs them by saying, *"One in (out), two in (out),"* etc.

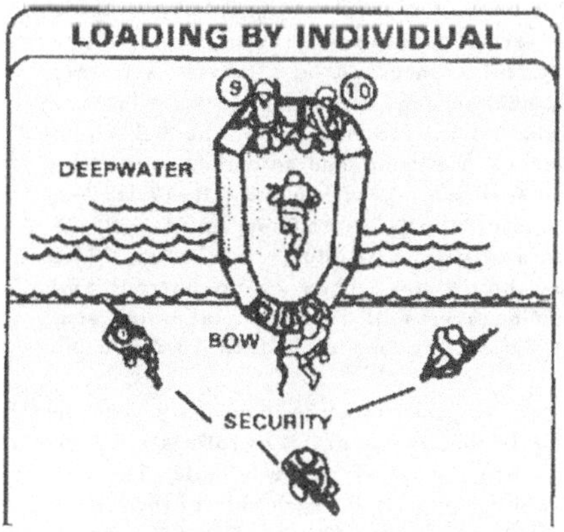

The crewmembers load and unload a boat in pairs when at shallow water riverbanks. They load or unload in pairs according to their number in the short count method. The coxswain directs them by saying, *"Ones in (out), twos in (out),"* etc.

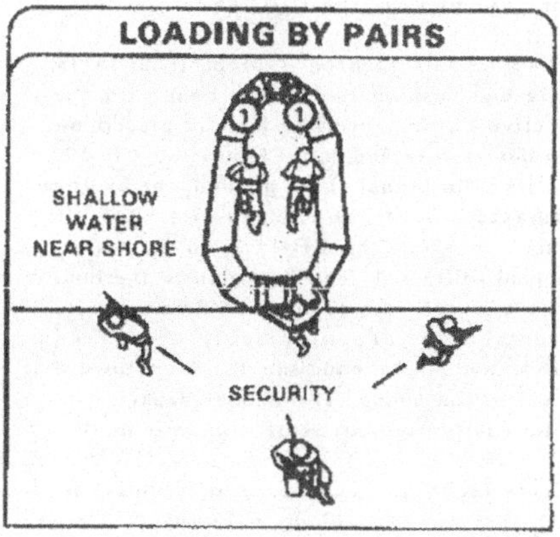

HELOCASTING

Helocasting is an excellent method for deploying troops and equipment in any terrain in which water courses exist. This technique involves a CH-47 helicopter, a 15-man rubber boat (RB-15), and a squad. The RB-15 is loaded with the squad's rucksacks, crew-served weapons, radios, and other heavy mission-essential items. All this equipment is kept in the RB-15 by a lashing system. At the desired time, the RB-15 is pushed off the ramp and into the water. The squad follows it, exiting the CH-47 in two columns off the tailgate. Drop speed is 20-25 knots and drop altitude is 10-20 feet. The water should have little or no current and should be free of all obstacles, including sea-weed and stumps, and be at least 15 feet deep.

The only preparation necessary for the CH-47 is that two lengths of rollers must be installed in the center of the tailgate. The two rearward set of seats on each side of the CH-47 must be raised to fit the RB-15 inside. Personnel wear fatigues (boots unbloused, shirts out, sleeves rolled down, top button fastened), load-carrying equipment, and an individual life preserver. The individual weapon is tied to the individual in such a manner that it can be raised overhead when the individual exits the CH-47.

The RB-15 lashing system is used in rigging and lashing the rubber boat with the respective equipment. The boat is placed on the roller system and moved into the CH-47, bow first. On signal, it is pushed out by the castmaster and coxswain or RB-15 commander. Once the RB-15 is in the water, the squad follows it. The first man to the boat makes a quick inspection for damage and accountability of equipment, frees the paddles, and starts paddling the boat toward the rest of the squad. If the boat capsizes, it can be easily righted using standard drills.

This technique can be used without the RB-15 to cast scouts along a riverbank. They would swim to a designated shore or to a designated point to conduct their mission.

RB-15 helocasting is also an effective means of resupplying a company operating along a water obstacle. Over 1,000 pounds of rations, ammunition, and supplies can be placed in each boat.

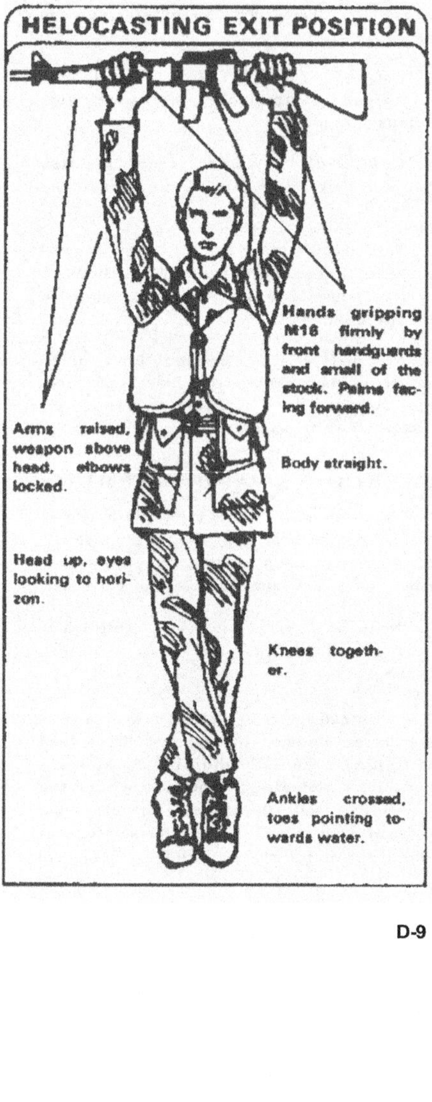

RIVER MOVEMENT

RIVER CHARACTERISTICS	
BEND 	A bend is a turn in the river course.
REACH 	A reach is a straight portion of river between two bends.
SLOUGH 	A slough (pronounced slue) is a dead end branch from a river, It is normally quite deep and can be distinguished from the true river by its lack of current.
DEAD WATER 	Dead water is a part of the river, due to erosion and changes in the river course, that has no currents. Dead water is characterized by excessive snags and debris.
ISLAND 	An island is usually a tear-shaped mass of land in the main current of the river. Upstream portions of islands usually catch debris and should be avoided.
CURRENT 	The current in a narrow part of a reach is normally faster than in a wider portion. The current is fastest on the outside of a curve; the sandbars and shallow water are found on the inside of the curve.
SANDBARS 	Sandbars are located at those points where a tributary feeds into the main body of a river or stream.

NAVIGATION

There are two acceptable methods of river navigation.

Checkpoint and General Route Method This method is used when the landing site is marked by a well-defined terrain feature and the waterway does not have many branches and tributaries. The navigator uses a strip map, with the route drawn on it, and looks for prominent checkpoints along the way. It is best used during daylight hours and for short distances. Except for those periods when the navigator is right at a checkpoint, this method is not completely accurate. It is, however, the easiest means of river navigation.

STRIP MAP

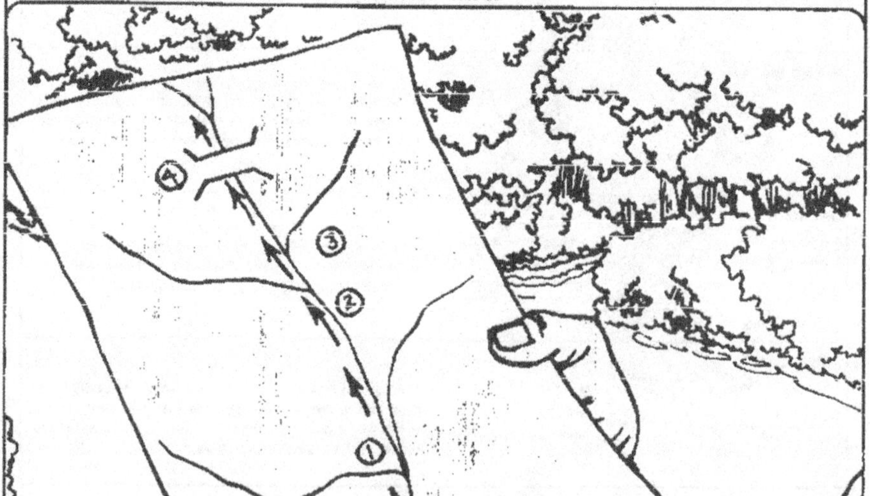

Navigator-Observer Method, This is the most accurate means of river navigation and can be used effectively in all light conditions.

The navigator is positioned in the center of the boat and does not paddle. During hours of darkness, he uses his flashlight under a poncho to check his map.

The navigator keeps his map and compass oriented at all times.

The navigator keeps the observer informed of the configuration of the river by announcing bends, sloughs, reaches, and stream junctions as shown on his map.

The observer compares this information with the bends, sloughs, reaches, and stream junctions he actually sees. When these are confirmed the navigator notes the boat's location on his map.

The navigator also keeps the observer informed of the general azimuths of reaches as shown on his map. The observer confirms these with actual compass readings.

The navigator announces only one configuration at a time to the observer and does not announce another until the first is confirmed and noted.

At night, a strip map drawn on clear acetate backed with luminous tape may be used instead of a map. It should be to scale or a schematic. It should show all curves and the azimuth and distance of all reaches. It should also show terrain features, streams, junctions, and sloughs.

TACTICAL MOVEMENT

The techniques of tactical river movement are very similar to those employed on land. As on land, movement techniques depend primarily on the likelihood of enemy contact and must be based on the concepts of traveling, traveling overwatch, and bounding overwatch.

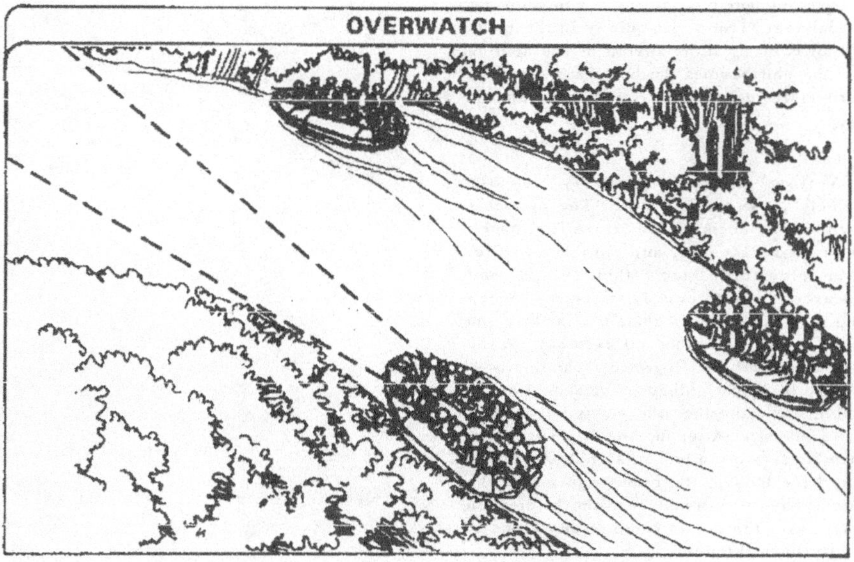

Distances between elements will vary depending upon observation, range of weapons, and means of communication used. Each boat must maintain visual contact with the boat to its front.

Boats move close to the shoreline, taking advantage of the natural concealment.

When bends in the river deny observation, a unit sends a reconnaissance team ashore to reconnoiter the river beyond the bend. When the reconnaissance team determines that the area is clear, it signals the boats to move forward.

Troops in the boats are assigned specific sectors in which to observe and fire.

One person is appointed in each boat as an air guard. If an enemy aircraft is sighted, the boats immediately move close to shore for concealment. Troops sit quietly in the boats until all is clear. If the aircraft makes a firing pass, the unit beaches the boats and takes the appropriate defensive actions for an air attack.

Actions taken on enemy contact resemble those taken ashore. The elements caught in the enemy's fire return fire, beach the boats, seek cover, and continue to fire. Other elements beach their boats and maneuver ashore to destroy the enemy. When a patrol is inserted by boat, the landing site must be secured before all elements of the patrol disembark. A suggested technique is to have the lead boat unload its personnel at the landing site while the other boats cover them from a distance. After the site is secured, the other boats are landed on signal. After the boats have landed, the crews either hide the boats or have them removed from the area. In either case, the crews remove any signs of activity on the landing site.

CAPSIZE PROCEDURE

RIGHTING BOAT

During small boat operations, there is always a chance of a boat being capsized (overturned) unexpectedly. In order for a unit to continue its mission, it must right the boat. To do this, the unit must follow this procedure:

The coxswain must first account for his men. He does this by having them count-off using the long count method. This will tell him who, if anyone, is injured or missing.

He next has the men hold on to the boat. They then pass their paddles to one man (designated by the coxswain before the operation). This man secures the paddles while the others right the boat.

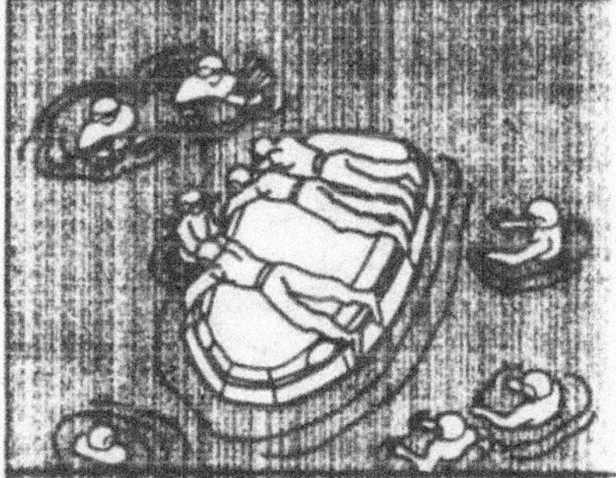

Next, three men are helped onto the boat. These men (all on the same side of the boat) grasp the bowline (which has been secured to one side of the boat), stand up, and prepare to pull the boat over. All but one of the men remaining in the water release the boat and move out to about 3 meters from it. The one man (in the water) holding on prepares to be pulled into the boat once the boat is pulled over.

The three men on the boat lean backwards and pull the boat over (upright). This also pulls the one man (in the water) into the boat.

Once the boat is upright, all of the men move to it. The one man in the boat helps the others into the boat. The man with the paddles passes them up to the other crewmembers and then climbs into the boat.

Once the crewmembers are in their positions, the coxswain again has the men count-off using the long count method. The men also check to see that their equipment is accounted for. The coxswain then gives the crew the appropriate orders for continuing the mission.

INTENTIONAL CAPSIZE

There may be times when a unit will need to capsize the boat intentionally. This may be necessary when the boat is full of water due to rough seas or heavy rainfall. To capsize the boat, the unit follows this procedure:

The coxswain must first have the men insure that all equipment is secured to the boat. He then gives the command, "**Pass paddles.**" The men then pass their paddles to one man—normally the man to either the left or right side of the coxswain. The coxswain also gives the command, "**Pass the bowline.**" The men pass the bowline and secure it along one side of the boat and in the rear.

The coxswain designates three men to remain in the boat. (They will capsize the boat after the others are in the water.) He then orders the other members out of the boat. He does this by giving the commands, "**Ones out, twos out,**" etc., until only three men remain in the boat. Once out of the boat, the men move out to about 3 meters from it.

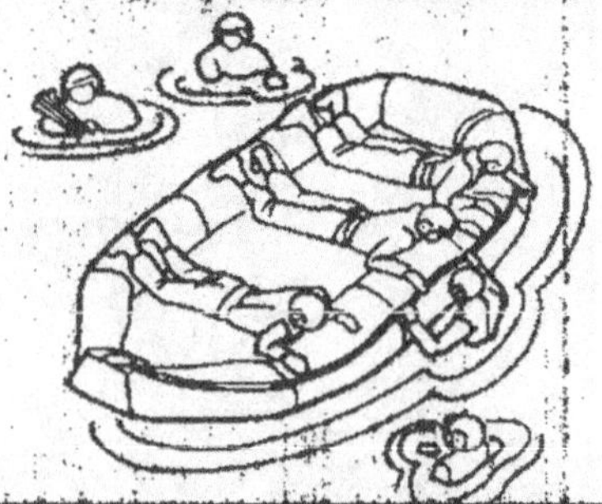

The coxswain then designates one of the men in the water to hold onto the boat in order to be pulled over onto the boat once it is capsized. The three men in the boat grasp the bowline, stand up, and lean backwards until the boat is capsized. This will pull the one man onto the boat when it is capsized.

The man now on top of the boat helps two other men onto the boat to help in righting it. They then follow the procedure outlined previously for righting a boat and continue the mission.

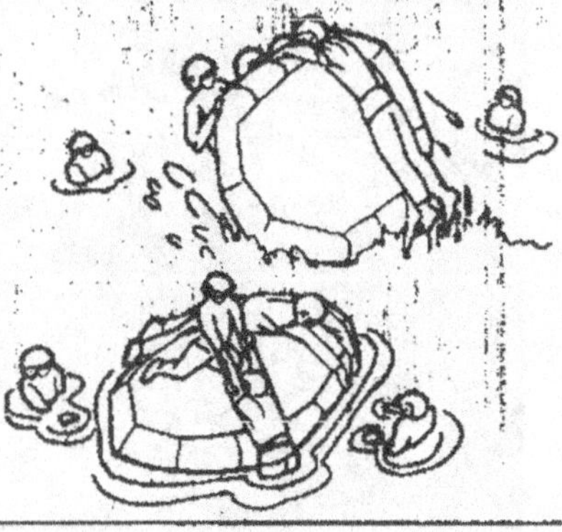

APPENDIX E

Fighting Positions and Shelters

Section I. GENERAL

Although jungles have excellent concealment, the jungle fighter must still prepare fighting positions to get as much cover as possible. Although not as important as fighting positions, jungle shelters provide shelter from the elements and make life in the jungle more comfortable and healthful.

Section II. PREPARING POSITIONS AND SHELTERS

FIGHTING POSITIONS

A jungle fighting position should provide cover, permit effective engagement of the enemy, and blend in with its surroundings. Swamps and low areas should be avoided because they normally do not afford good fields of fire and are too wet to permit the digging of fighting positions.

PROTECTION FROM ENEMY FIRE

A good fighting position provides protection from direct small-arms fire and indirect fire fragmentation. Protection from small-arms fire is provided by frontal cover. Natural frontal cover (trees, rocks, logs, rubble, etc.) is best as long as it is strong enough to stop bullets. Natural cover blends with the surroundings; therefore, it is hard for the enemy to identify positions. It maybe necessary, however, to use the dirt from the hole to build frontal cover when natural frontal cover is not available. Frontal cover must be wide enough to provide room for two soldiers, thick enough (at least 46 centimeters [18 inches] of dirt) to stop enemy small-arms

CONTENTS

fire, high enough to cover the heads of the soldiers firing from behind it, and far enough in front of the hole to provide room for elbow rests and sector stakes.

Protection from indirect fire fragmentation is gained by getting the soldiers below ground level and by building overhead cover. The logs cut to build overhead cover must be strong. Soft wood will collapse when struck by shell fragments.

To insure complete protection for fighting positions, it is necessary to build flank and rear cover. This cover protects against the effects of indirect fire that bursts to the flanks or rear of the position and against the effects of friendly supporting weapons located in the rear.

To conceal the fighting position, troops should take advantage of the jungle's natural cover. They should use large fallen or standing trees and depressions in the ground, and work to improve upon the natural cover provided. Some logs, such as palm tree logs, are too soft to stop bullets. If forced to use soft wood in building cover, a 10-inch layer of dirt or sandbags will have to be added to the protection provided by the wood.

FIGHTING POSITION

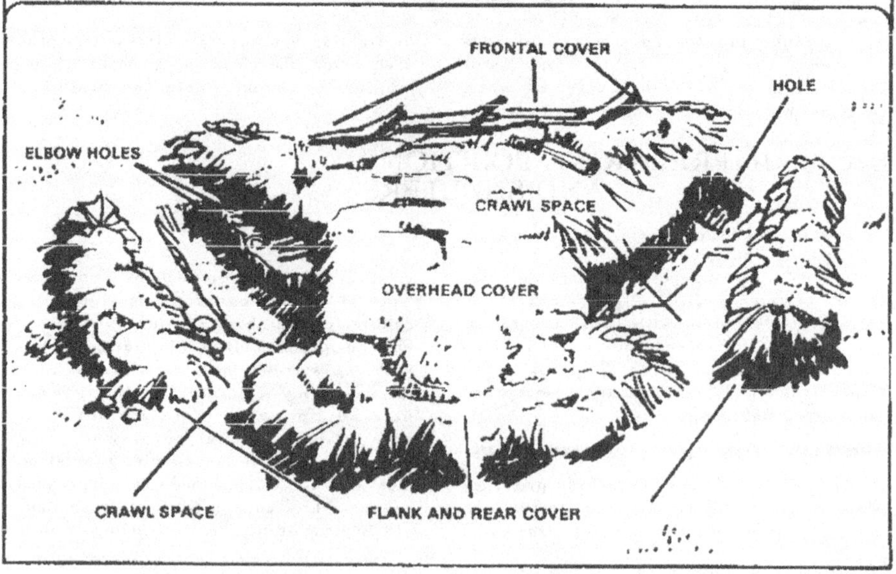

When positions are constructed, all artificial cover must be camouflaged to blend in with the surrounding terrain so it cannot be detected by the enemy. Camouflage should be kept fresh-dead or rotting vegetation will give away positions. Troops can transplant live plants—the moist soil will help these plants survive.

All positions should provide for drainage during tropical rainstorms. The floor of the position should slope from the center of the position toward the grenade sumps. This is important during the rainy season to avoid immersion foot.

If time permits, an elevated floor stand of saplings can be built so that soldiers do not have to stand in mud or water. Grenade sumps must remain clear.

Jungle soils are soft and erode quickly during the rainy season. Revetments will prevent the walls of a position from caving in. The stakes should be driven all the way into the ground, once the anchor lines are attached. They must then be concealed so that they cannot be confused with the firing stakes.

OBSERVATION AND FIELD OF FIRE

Soldiers should clear away only what is absolutely necessary, in order to retain as much natural concealment as possible. In properly constructed positions, the soldier's eyes and weapon will be at ground level. In this way, the soldier is concealed from the enemy, is protected by cover, but still can detect any approaching enemy. If a small tree is in a soldier's field of fire, it should not be chopped down completely. Only the branches that deny observation should be removed. The cuttings on the tree should be darkened with mud. Overly cleared areas or fresh, improperly concealed cuttings are easy to spot.

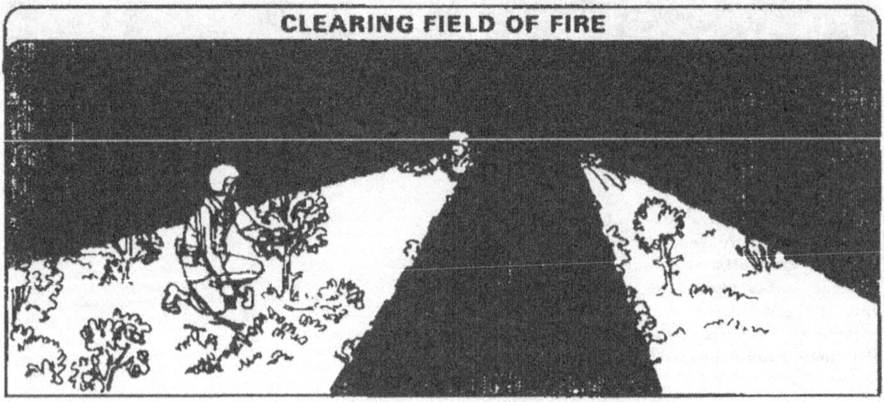

CONCEALMENT FROM ENEMY AIR OBSERVATION

Positions should be placed under the jungle canopy for its natural concealment from air observation. If this is not possible, positions should be covered with camouflage nets or with expedient covers of loosely woven vines and branches.

For more detailed information on how to prepare jungle fighting positions, *see FM 7-8.*

SHELTERS

Shelters are made when the tactical situation permits. The poncho can be used as a roof for the shelter. It reduces the need for extensive concealing foliage and provides protection from the elements. However, natural camouflage must still be applied to break up the outline of the poncho and reduce the glare of a wet poncho.

EXAMPLES OF PONCHO SHELTERS

HASTY SHELTER-CANOPY FASHION

A hasty shelter is made by suspending the poncho from low underbrush. Due to its simplicity, it can be easily erected at night, especially if heavy strings have already been tied to the corners of the poncho.

HASTY SHELTER-CANOPY FASHION

This is another hasty shelter pitched canopy fashion.

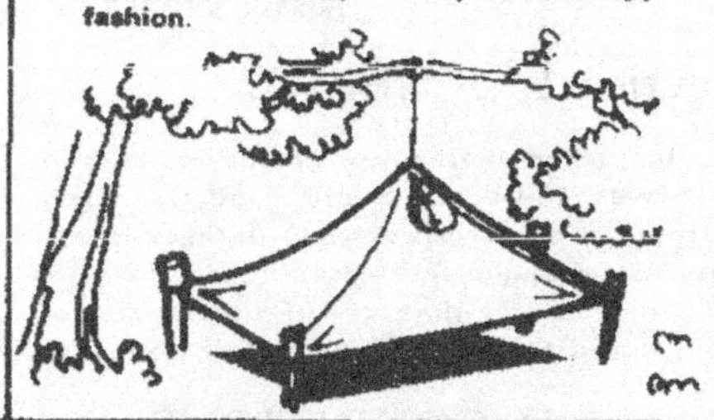

PONCHO AND SPREADER BARS

This is a hasty shelter using a poncho and two branches for spreader bars.

LOW SILHOUETTE SHELTER

This low silhouette shelter can be used while improving fighting positions. It can be lowered by removing the front upright supports.

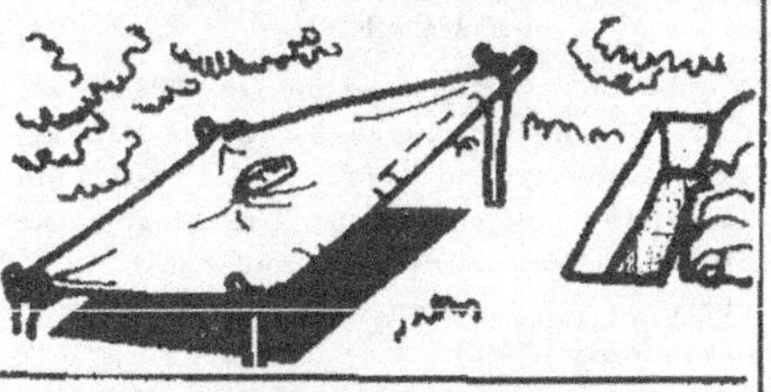

PONCHO SHELTER

Two ponchos fastened together will shelter four soldiers from the rain. Extra ponchos can be used as ground sheets.

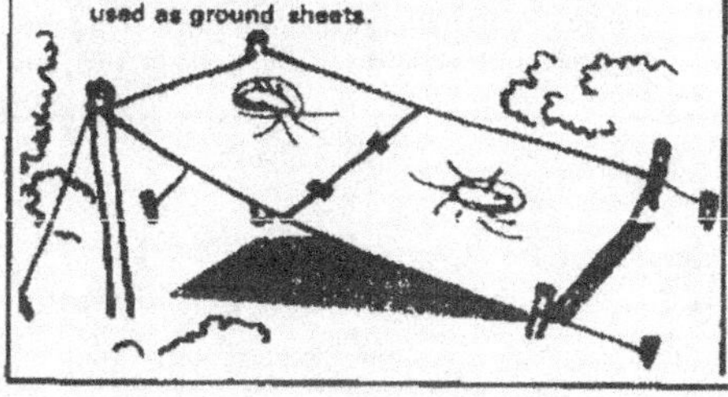

SLEEPING-PLATFORM AND FOOTREST

The following type of shelter may be used for a longer stay in more secure areas. A sleeping-platform and footrest protect from dampness and insects.

APPENDIX F

NBC Warfare in Jungle Areas

Section I. GENERAL

Jungle areas require exceptionally high standards of discipline and conditioning to maintain an effective NBC defense readiness. The demand for these higher standards results from the jungle's high temperature and humidity and heavy rainfall. Sunlight and wind are reduced within thick jungles due to the vegetation. These conditions affect NBC operations because they reduce the chances that NBC agents will disperse, thus increasing their impact on operations in that area.

Section II. CONDITIONS CREATED BY NBC AGENTS

SPECIAL CONDITIONS

NUCLEAR

The initial effects of nuclear blasts are not significantly reduced by jungle foliage. The canopy may provide some protection against thermal radiation. The blast wave will blow down many trees and cause flying debris. These fallen trees may restrict movement, but may also improve observation and fields of fire of some weapons. Fallout may be temporarily retained in the jungle canopy, reducing the immediate hazard. Later rains, however, will wash these particles to the ground and concentrate them in low areas. These areas are likely to become radiation "hotspots."

BIOLOGICAL

Jungle climates favor the use of biological agents. Biological agents thrive in heat, humidity, and shade. As with chemical agents, downwind and spray hazards will be reduced by the lack of wind in the jungle. Strict supervision of field sanitation and adequate purified water supplies will provide the best defense against biological agents as well as control natural diseases in the jungle.

CONTENTS

CHEMICAL

If jungle enemies use chemical weapons, they can cause serious problems for US troops. Persistent agents delivered by means of artillery shells will remain effective for long periods under the canopy. The jungle heat will vaporize chemical agents that are delivered in liquid form. Because of low wind speeds, these vapors will hang in the air and remain hazardous for long periods. The downwind hazard, however, will be reduced. Likewise, the danger from sprayed agents will be reduced. Protective masks and clothing will be uncomfortable in high heat and humidity. Commanders must plan for a decrease in performance by soldiers in this environment. They should also plan for heat casualties. Finally, NBC equipment must be maintained so that it does not rot, mildew, or rust.

Section III. ACTIONS TAKEN TO PROTECT TROOPS

JUNGLE MISSION ORIENTED PROTECTION POSTURE

When temperatures rise to the 85 to 100 degree Fahrenheit range, troops can be expected to continue moderate or heavy workloads only if they are permitted to reduce their mission oriented protection posture (MOPP) level. This increases the risk of chemical casualties if the unit comes under chemical attack. Vision, especially downward vision, is limited while wearing the mask. While these factors tend to discourage the wearing of protective equipment, leaders must balance the need to accomplish their mission as opposed to the NBC threat.

As in any other environment, the commander must plan his MOPP level based upon the mission and the estimate of the chemical threat. The level and the length of exertion are important factors. Long-term efforts, such as a road march, are more taxing than short-term, high-energy tasks, such as an assault.

A commander considering protective measures should keep two principles in mind. The first is that although troop safety is an important consideration, the accomplishment of the mission is of greater importance. The second is that even though a higher level of protection increases the risk of heat casualties, these casualties are not as hazardous or as long lasting as are chemical casualties. Heatstroke, for example, is fatal about 50 percent of the time, but makes up a very small percentage of all heat casualties. Physical conditioning and acclimatization of soldiers have an impact on how long they can operate in protective equipment.

MOPP

MOPP	PROTECTIVE EQUIPMENT			
	OVERGARMENT	BOOTIES	MASK/HOOD	GLOVES
1 (POSSIBLE)	Worn, opened or closed based on temperature	Carried	Carried	Carried
2 (POSSIBLE)	Same as MOPP 1	Worn	Carried	Carried
3 (LIKELY)	Same as MOPP 1	Worn	Worn, hood opened or closed based on temperature	Carried
4 (IMMINENT)	Worn, closed	Worn	Worn	Worn

Assuming troops are acclimatized, work rates can be exceeded for short periods if adequate rest and water are provided. In all cases leaders must balance the chances of exposing their units to heat exhaustion or chemical agents to the need to accomplish the mission. The ideal MOPP is the category that provides the highest degree of protection from chemical effects and still permits the mission to be accomplished.

The following measures can improve unit efficiency at any MOPP level:

- Rotating heavy work among elements or individuals
- Authorizing longer and more frequent rest periods
- Providing adequate drinking water
- Using truck or air transport when possible

NOTE: For further information, see FM 21-40.

APPENDIX G

Communication Techniques in a Jungle Environment

Section I. GENERAL

Rapid, reliable communications are essential in jungle operations. Command, control, fire support, resupply, and evacuation are all dependent on effective communications. The importance of establishing and maintaining reliable communications cannot be overemphasized. For this reason, a commander must give communications high priority in his planning and supervision.

Jungle operations place additional demands on the resources of communications units and personnel. Tactical and signal units may require augmentation of signal personnel and equipment to accomplish the command and control functions. Besides providing normal tactical communications networks, signal units may have to operate long distance radio sets. Army aircraft will have to be used in many instances to provide a radio relay capability as well as to assist wire-laying operations.

Section II. COMMUNICATING IN THE JUNGLE.

TACTICAL COMMUNICATIONS MEANS

With the range of FM communications in a jungle environment significantly reduced due to the dense undergrowth, heavy rains, and hilly terrain, all means of communications should be used in a manner that complements one another. For all operations, backup means of communications must be planned. The common means of communication available are visual, sound, messenger, wire, and radio.

CONTENTS

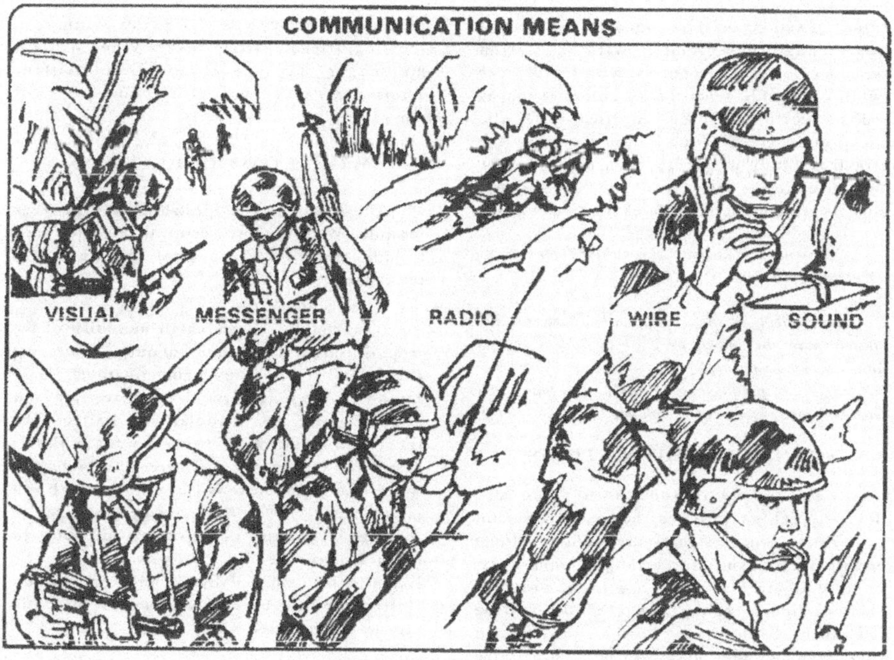

VISUAL COMMUNICATIONS

Such means as arm-and-hand signals, pyrotechnics, flashlights, headlights, smoke grenades, mirrors, and panels normally allow quick transmission of messages and instructions. However, visual means of communication in the jungle are restricted by the dense vegetation. Commanders will rarely have visual contact with all elements or members of their unit.

Arm-and-hand signals are used in all types of operations. The effectiveness of arm-and-hand signals can be improved by insuring that each soldier understands the meaning of the signal and passes it on to others in his vicinity.

Pyrotechnics can be used in most conditions of visibility. Certain standard colors of smoke or flare signals have limited use because they blend in with vegetation (green) or offer little contrast with fog or haze (white). Careful selection of colors (such as red or yellow) which contrast with the background color of the jungle increases the effective range of pyrotechnic signals. The jungle canopy can affect the use of star clusters and parachute flares in two ways.

Overhead clearance required for firing flares, is limited,

Once clear of the canopy, the flare or star cluster may be hard to see by other ground units looking through the canopy. The disadvantage of pyrotechnic signals is that they can be easily seen and imitated by the enemy.

SOUND COMMUNICATIONS

These means include both voice and devices such as whistles, horns, gongs, and explosives. Sound signals are used to attract attention, transmit prearranged messages, or spread alarms. They are good only for short distances. Range and reliability will be reduced by battle noise, weather, terrain, and vegetation. As they may also be heard by the enemy, they should be restricted for security reasons. Sound signals must be simple to avoid misunderstandings. The means for sound signals are usually prescribed by the unit SOP and communications-electronics operation instructions (CEOI).

MESSENGER COMMUNICATIONS

These types are most secure and a good way to send long messages and documents. However, the speed of surface messengers is severely limited by jungle terrain and vegetation. The lack of roads restricts full use of motor vehicles. Foot messengers may have difficulties with land navigation and natural obstacles. The use of air messengers maybe a better alternative, but one that is limited by availability of aircraft, scarcity of good landing zones, weather conditions, and the fact that air activity aids the enemy in locating friendly troop areas. When using a messenger, messages should be written. Their text must be clear, concise, and complete.

WIRE COMMUNICATIONS

These methods are established whenever feasible. Wire is more secure than radio, hard to jam, and allows conversation with break-in capability.

The heavy rain and high humidity of the tropics reduce the range (about 20 percent) and reliability of wire communications. Wire laying in the jungle requires more time and suitable w-ire laying routes are limited. The increased cover and concealment make it easier for the enemy to intercept or interrupt wire communications and to ambush maintenance crews. Wire laid by air on top of the jungle canopy is extremely difficult to maintain and recover; however, it is less likely to be damaged by vehicles or weather. Splices should be carefully made to avoid signal losses when wet.

RADIO COMMUNICATIONS

Radio communications are normally the fastest means available but the least secure.

Effective radio communications in the jungle require emphasis on:

- **Use of antennas and field expedients**
- **Use of retransmissions and radio relays**
- **Continuous preventive maintenance**
- **Communications security (CO MS EC)**
- **Trained operators**

The usefulness of radio communications is reduced in jungle operations. VHF and UHF radios (FM radios included) are particularly limited by the jungle growth and terrain that absorb and obstruct the transmissions. It is not unusual for the range of a set operated in the jungle to be reduced by 10 to 25 percent of the normal range.

ANTENNAS AND FIELD EXPEDIENT ANTENNAS

In order to overcome environmental conditions in the jungle, existing antennas must be used properly and field expedient antennas produced to extend the capacity of current equipment.

> To use existing antennas properly, the following items must be accomplished:
>
> - Keep whip antenna vertical when transmitting
> - Insure the antenna is not grounded by being in contact with foreign object
> - Position antennas to achieve the best line-of-sight possible between stations (such as on top of a hill)

The problems of line-of-sight antenna locations and operations security are best solved by remoting the transmitter from the command post.

The AN/GRA-39 radio set control group allows the operator to remote the radio up to 3.2 kilometers away from the observation post. It should be noted that there is no practical advantage to be gained by installing the radio and antenna away from the command post (remoting) unless the distance is 1 kilometer or more. The error inherent in radio direction finding equipment does not provide for a consistent accuracy less than 1 kilometer.

REMOTE SETUP

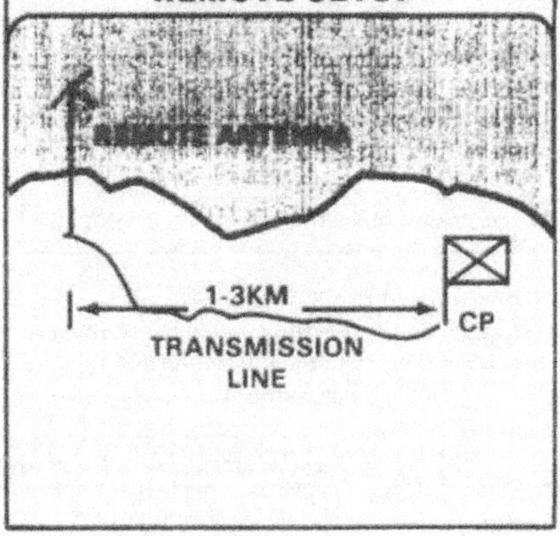

Field expedient antennas will enhance the communications capacity of a unit operating in the jungle if they are properly constructed. WD-1 or portions of the RC-292 antenna can be used as the radiating element (antenna). Wood or plastic spoons can be used as insulators when insulators are not available. Resistors should be obtained through communications maintenance channels.

> Useful field expedient antennas in the jungle are:
>
> Field expedient RC-292
>
> Patrol antenna

The standard issue RC-292 is a highly-effective, omnidirectional antenna. It is usually more effective than a whip antenna, and is particularly effective in the VHF range. With all its components, it weighs 48 pounds and two soldiers can erect it in 15 minutes.

Leaders should consider using field expedient versions of this antenna in the jungle for these reasons:

- Excessive load for dismounted soldiers
- Awkward to assemble in restricted terrain
- Increased chance of losing parts

Portions of the RC-292 may be used to construct a field expedient antenna. Only the vertical sections, antenna base, ground plane sections, and transmission line are needed.

FIELD EXPEDIENT ANTENNA USING PORTIONS OF RC-292

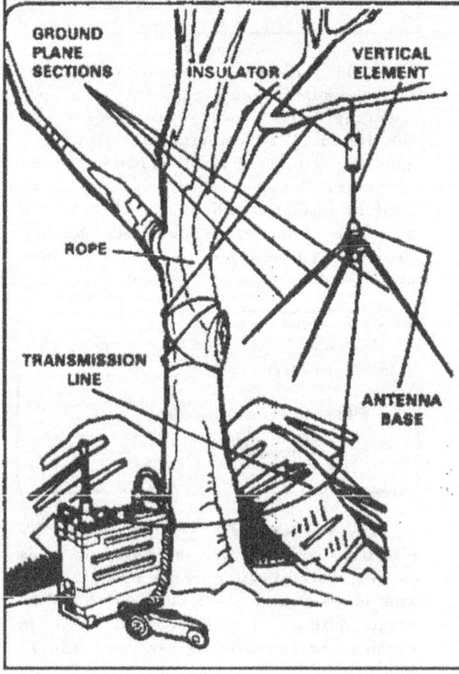

To construct this antenna:

Attach an insulator to the vertical element and add a piece of rope to it.

Tie a weight on the rope and throw it over a tree limb.

Pull the antenna up as high as necessary.

Raise or lower the antenna to change element lengths each time the frequency is changed.

Attach the coaxial lead-in to the radio connector as with the standard RC-292.

Add a rope from the bottom of the head to a stake in the ground. This prevents swaying and damage to the equipment when there are strong winds.

Another field expedient antenna can be constructed without using any of the standard RC-292 components. This antenna has various names such as the field expedient 292, the jungle antenna, and the field expedient ground plane antenna.

FIELD EXPEDIENT 292 ANTENNA

Construction is as follows:

Step 1. Using the quick reference chart at the end of this appendix, cut all four wires for a quarter wave antenna. Connect them as in the following illustration.

Step 2. Two insulators are needed; one at each end to separate the vertical element. Attach a rope with a rock tied to it to throw the rope over a tree limb.

Step 3. Connect the WD-1 as shown, before pulling the antenna up in the air.

Step 4. At the radio, connect the vertical element wire of WD-1 to the antenna connector and the ground plane wire to the radio set chassis.

If frequencies are known before deployment, soldiers can cut wires to the appropriate lengths and carry them on the operation.

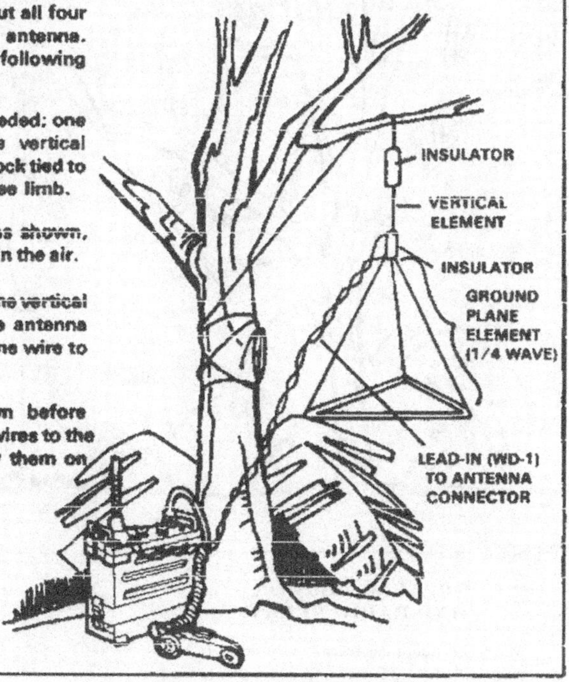

A third type of expedient antenna used in the jungle is the *patrol antenna.* It is the same as the jungle antenna minus the ground plane element. Insure that the length is determined by using the 1/2 wavelength instead of the 1/4 wavelength column of the chart. The range of the set will be increased 2 to 3 times its normal range. This antenna is easier to handle and construct.

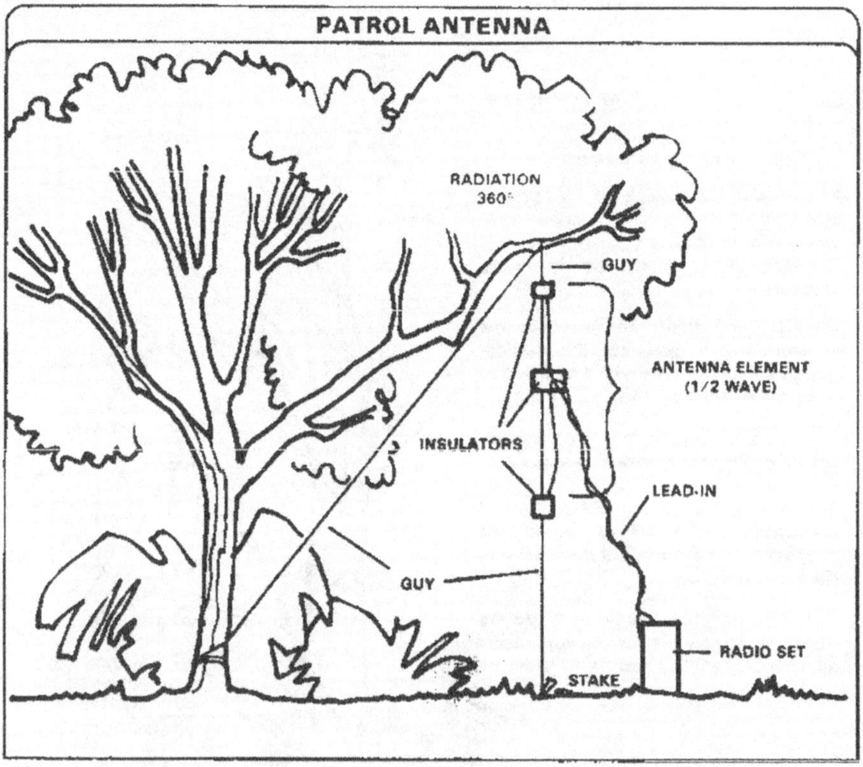

RETRANSMISSION AND RADIO RELAY

With the decrease in range of most radios used in jungle operations, retransmission will be a good method to improve communications. Commanders should consider establishing retransmission sites and using additional signal personnel. Radio relay sites, either airborne or on high terrain, will also improve radio communications between stations.

EQUIPMENT MAINTENANCE IN THE JUNGLE

Moisture commonly found in the jungle environment can result in numerous outages of communications-electronic equipment.

The following measures, combined with normal maintenance, will help counteract the problems.

- Seal the radios with silicone compound.
- Pack radios in waterproof containers when not in use (remove batteries).
- Protect handsets and microphones with plastic bags. (Batteries and rations are packed in plastic bags.) Check frequently for moisture build-up in the bags.

When radios are protected by plastic bags, make sure the battery vent is not obstructed. Radios protected in this manner must be constantly checked to insure moisture does not build up in the bag. Clean radios as frequently as individual and crew-served weapons are cleaned.

Section III. AVOIDING ENEMY INTERFERENCE

COMMUNICATIONS SECURITY

In the past, jungle enemies have relied heavily on friendly radio messages as primary intelligence sources. Communications security denies or delays unauthorized persons from gaining information from telecommunications.

ROLE OF COMSEC

It includes:

- Using authentication to insure that the other communicating station is a friendly one.
- Using only approved codes.
- Designating periods when all radio equipment is turned off.
- Restricting the use of radio transmitters (monitoring radio receivers/listening silence).
- Enforcing net discipline and radiotelephone procedures (all stations in a net must use authorized callsigns and prowords; and they must limit transmissions to official traffic).
- Selecting radio sites with a hill as a shield between them and the enemy.
- Using directional antennas when possible.

GUIDELINES FOR BETTER COMMUNICATIONS SECURITY

- Enforce proper use of authentication systems.
- Use only authorized authentication systems.
- Assume that your radio transmissions are being intercepted by the enemy.
- Use proper site selection, minimum power, and minimum transmission times.
- Do not use electrical communications when messengers will do as well. Messengers are the most secure means of communications, other than personal contact.
- Do not paraphrase or try to talk around the subject. Keep transmissions short and concise. Use authorized codes to send information.

ANTIJAMMING

Radio operators should use the following antijamming procedures to thwart enemy jamming efforts.

Recognition. The first thing an operator must do when there is interference on his radio is to try to find its cause. As symptoms ofjamming are often similar to other types of interference, he should not assume that it is jamming. If the interference decreases when the receiver antenna is removed, the interference is jamming; if it does not, the interference is generated inside the receiver.

Continued Operation. Normal radio operation is continued once jamming has been identified so that the enemy cannot determine the effect of his jamming. The rule is: during jamming, operators continue operating unless ordered to shut down.

Reporting. All operators must report jamming to their next higher headquarters, by another means of communications; for example, wire or messenger. A typical jamming report tells *date and time of jamming, frequencies affected, type and strength of jamming signal, and designa tion of the unit making the report.*

The exact format of the report is found in the unit's CEOI.

FORMULAS FOR COMPUTING WAVELENGTH

TO FIGURE:	FORMULA:	EXAMPLE:
A quarter wave-length in feet:	Divide 234 (constant) by the operating frequency in MHz.	234 ÷ 44.8 = 5.22' or 5'3".
A half wave-length in feet:	Divide 468 (constant) by the operating frequency in MHz.	468÷56 = 8.36' or 8'5".
A full wave-length in feet:	Divide 936 (constant) by the operating frequency in MHz.	936÷45 = 20.8' or 20'10".
TO CONVERT:		
Feet to meters:	Multiply by .3048 (constant).	110' x .3048 = 33.5 meters.
Meters to feet:	Multiply by 3.28 (constant).	100 meters x 3.28 = 328 feet.

QUICK REFERENCE CHART EXPEDIENT ANTENNAS

HIGH FREQUENCY (HF) ANTENNA LENGTH IN FEET AND INCHES				VERY HIGH FREQUENCY (VHF) ANTENNA LENGTH IN FEET AND INCHES			
Op freq in MHz	1/4 wave	1/2 wave	1 wave	Op freq in MHz	1/4 wave	1/2 wave	1 wave
2	117′	234′	468′	30	7′10″	15′7″	31′2″
3	78′	156′	312′	33	7′1″	14′2″	28′4″
4	58′6″	117′	234′	35	6′9″	13′5″	26′10″
5	46′9″	93′7″	187′4″	37	6′4″	12′7″	25′2″
6	39′	78′	156′	40	5′10″	11′8″	23′4″
7	33′5″	66′10″	133′8″	43	5′5″	10′10″	21′8″
8	29′3″	58′6″	117′	45	5′3″	10′5″	20′10″
9	26′	52′	104′	48	4′10″	9′8″	19′4″
10	23′5″	46′10″	93′8″	50	4′9″	9′5″	18′10″
11	21′3″	42′6″	85′	55	4′3″	8′6″	17′
12	19′6″	39′	78′	57	4′1″	8′2″	16′4″
13	18′	36′	72′	60	3′11″	7′10″	15′8″
14	16′9″	33′5″	66′10″	65	3′7″	7′2″	14′4″
15	15′7″	31′2″	62′4″	68	3′5″	6′10″	13′8″
16	14′7″	29′2″	58′4″	70	3′4″	6′7″	13′2″
17	13′9″	27′6″	55′	75	3′1″	6′2″	12′4″
18	13′	26′	52′	80	3′	5′11″	11′10″

APPENDIX H

The Individual Soldier's Combat Load

Section I. GENERAL

All equipment considered essential to mission completion should be carried by the individual at all times. Rucksacks should be packed in such a way that they can be dropped or hidden, for recovery later, to allow troops to move freely upon contact or on patrol. In an emergency, they may be discarded completely. Those essential items that must be placed in the rucksack, due to the extended nature of operations, should be removed when the situation dictates. Combat-essential items should not be left in rucksacks that are cached or dropped. Unit SOP should dictate the items that are removed from rucksacks and included in equipment carried by each soldier when on short-duration missions where rucksacks are left behind. Packing lists should also be designed for squads and platoons with careful attention given to balancing loads. (For example, machinegunners should not be given extra equipment to pack until the fighting and existence loads of the other members of the platoon are equal to that of the machinegunners.)

Section II. PREPARING THE LOAD

WEIGHT CARRIED

The problem of excessive weight in existence loads needs to be closely monitored by commanders at all levels. Specific company and platoon SOPs need to be reinforced with inspections prior to deployment, demanding strict adherence to packing lists. Unusable equipment should not be carried.

WATER CARRIED

Water will constitute the second greatest part of each soldier's load. Jungle fighters must be issued more canteens than normal, and they may have to carry 5 to 10 quarts of water attached to their rucksacks, especially in the dry season. Water purification tablets should be carried so that stream water may be used.

AMMUNITION CARRIED

Units should be told how long they can expect to operate without resupply. This figure becomes the basis for planning the equipment, supplies, and ammunition to be carried. Ammunition will normally make up the greatest part of each soldier's load, and ammunition supplies must be planned to last until troops can be resupplied. Rifle and machinegun ammunition and grenades should be loaded on pallets in a rear area so that units can easily be resupplied by helicopter, if required. Every soldier should carry colored smoke grenades and flares for signaling, and tripflares and Claymore mines for security. Special ammunition may be carried, if required by the mission. The most common types of special ammunition

CONTENTS

and C4 explosives, concussion grenades, and CS munitions.

The following description of a soldier's combat load should be used *only as a guide* in developing a unit SOP. Fighting and existence loads will vary according to each unit's special needs and missions.

FIGHTING LOAD FOR AN INFANTRY PLATOON CONDUCTING OPERATIONS IN JUNGLE TERRAIN

	INDIVIDUAL (Carried at all times)				SQUAD AND TEAM LEADERS (Additional)	ADDED FROM RUCKSACK (When rucksacks are left in patrol base or cached, or when on a mission where no rucksacks are carried.)	
	CLOTHING	EQUIPMENT/BASIC LOAD					
	Helmet, with liner, camouflage cover and band. Trousers and jacket (camouflage, jungle fatigue). Belt. Socks. Jungle boots. ID tags.	M16 rifle with sling and one magazine (30 rounds). Ammunition, 180 rounds and six magazines. The following weapon and basic loads, substituted as appropriate: M203 with one round of 40mm HE and one magazine of 5.56mm (30 rounds) Basic load 40mm: 36 rounds HE Basic load 5.56mm 140 rounds and magazine	M60 machine gun with sling Basic load 200 rounds per gun. Caliber .45 pistol for gunner with magazine, seven rounds Two magazines (14 rounds) with ammunition pouch. Ammunition cases (two for magazines). Hand grenades (two fragmentation). Smoke grenades (two). Canteens (two 1-quart canteens—filled). (Amount depends on availability of local water.)	Water purification tablets. Individual equipment belt/first-aid packet case and suspenders. Snaplink. Twelve-foot nylon rope (Swiss seat). Bayonet knife and scabbard. Ammunition case containing items such as: insect repellant, camouflage stick, foot powder, toilet paper, bandages, matches, notebook and pencil, aspirin, emergency rations, penlight. Individual cleaning equipment. Rations (one meal, C-rations).	Compass. Strobe light. Map and case. Flashlight with batteries. Whistle. First-aid kit, general purpose. Watch.	One poncho per man. One entrenching tool with carrier. Two machetes with case (two per squad). Mask, CBR M17A1 with carrier (optional depending on threat).	NOTE: Critical items such as Claymores, LAWs, M60 ammunition, etc., should never be left with rucksacks.
APPROX WEIGHT (POUNDS) TOTAL	9.36	+	42.63	= 51.99	5.06	8.17	

EXISTENCE LOAD FOR AN INFANTRY PLATOON CONDUCTING OPERATIONS IN JUNGLE TERRAIN

ON OR IN RUCKSACK			ADDITIONAL EQUIPMENT/BASIC LOADS (To be distributed through platoon during sustained operations. This equipment should first be broken down by squads, then distributed to individuals based on balancing loads across the board.)	
EQUIPMENT		RATIONS		
One frame including lower back, shoulder, and waist straps. Pack, medium, combat, field. Bag, waterproof. Poncho (prerigged for shelter). Poncho liner. Socks (four pairs). Trousers (camouflage). Tennis shoes. One sleeping shirt or tee shirt. Health items: soap, toothbrush, razor and blades (use insect repellent for shaving). Insect bar.	Air mattress. Extra insect repellent. Extra water purification tablets. Two quarts water. (Amount depends on availability of local water.) Twenty-foot suspension line. Sewing kit. Mask, CBR M17A1 with carrier (optional depending on threat). Entrenching tool. Sterno/heat tablets/butane stove. Life preserver, inflatable (optional).	C-rations (dry rations are lighter, but more water is needed). 1 day: 1.75 per meal x 3 number of meals = 5.25 TOTAL 3 days: 1.75 per meal x 9 number of meals = 15.75 TOTAL 5 days: 1.75 per meal x 15 number of meals = 26.25 TOTAL	Two PRC-77 radios, complete with four batteries. Five TA1 telephones, w/200 meters wire (one for plt HQ, one per squad). Twelve Claymore mines, four per squad. Twelve LAWs, four per squad. Two M60 machineguns with 600 rounds (200 rounds for gunner, 200 rounds for assistant gunner, 200 rounds for ammunition bearer). Fifteen parachute flares, five per squad.	Fifteen trip-flares, five per squad. Three 120-foot ropes (optional), one per squad. Eight machetes w/carrying case (two per squad, two for plt HQ). NOTE: This weight will increase as unit strength decreases. Radio telephone operators, M60 machinegunners, and grenadiers are left out to determine the average extra weight per soldier.

	WITHOUT OPTIONS	WITH OPTIONS	DAYS	1	3	5		
APPROX WEIGHT (POUNDS) TOTAL	29.07	38.54	-	34.32	44.82	55.32	195	(Approximately 6 pounds per rifleman.)

EXISTENCE ITEMS—RUCKSACK

	RIFLEMAN	APPROX WEIGHT (POUNDS)	M60 MACHINEGUNNER	APPROX WEIGHT (POUNDS)
RUCKSACK	As listed with options—air mattress and life preserver only.	34.57	As listed with options—air mattress and life preserver only.	34.57
RATION	Three days (note 1).	15.75	Three days (note 1).	15.75
ADDITIONAL BASIC LOAD	One Claymore mine (plt/sqd items—approximately 6 pounds per rifleman).	4.00		
	One parachute flare.	1.00		
	One trip flare.	1.00		
EXISTENCE LOAD		50.32		50.32
EXISTENCE LOAD TOTAL		56.32		50.32
COMBAT LOAD (NOTE 2) TOTAL		99.01		106.98

NOTE 1: Dry rations are lighter and more compact than C-rations. Water may be a problem, however.

NOTE 2: By adding total fighting load and total existence load, to obtain the total combat load, a commander can determine who can receive any additional items to carry.

SAMPLE INDIVIDUAL FIGHTING LOAD FOR 5-DAY OPERATION IN A TROPICAL ENVIRONMENT WITH NO NBC THREAT

RIFLEMAN	APPROX WEIGHT (POUNDS)	M60 MACHINEGUNNER	APPROX WEIGHT (POUNDS)
CLOTHING As listed.	9.36	CLOTHING: As listed.	9.36
EQUIPMENT		EQUIPMENT	
M16 w/mag (30 rounds)	7.91	M60	23.00
Ammunition (180 rounds)	6.21	200 rounds	14.00
Ammunition cases	.86	Cal .45 w/mag (7 rounds)	3.20
Hand grenades (2 frag)	2.00	Ammunition pouch (14 rounds)	.75
Smoke grenades	1.50	Canteens (two 1-quart—filled)	7.20
Canteen (two 1-quart—filled)	7.20	Water purification tablets	.06
Water purification tablets	.06	LBE	1.59
LBE	1.59	Snaplink	.25
Snaplink	.25	Twelve-foot nylon rope	.75
Twelve-foot nylon rope	.75	Ammunition case (subsistence items)	2.00
Ammunition case (subsistence items)	2.00	Cleaning equipment	.75
Gloves	.50	Rations	26.25
Cleaning equipment	.75		
Rations	26.25		
EQUIPMENT	57.83		79.80
CLOTHING	+ 9.36		+ 9.36
TOTAL FIGHTING LOAD	67.19		89.16

Section III. CARRYING THE LOAD

USE OF THE RUCKSACK

Platoon- and company-sized units conducting offensive operations in jungle terrain must emphasize self-reliance. Units must carry rucksacks and be able to operate for extended periods of time with routine resupply being accomplished every 5 to 7 days. This will allow units to develop their operational areas, and preclude compromising their location for frequent resupply operations.

To minimize the weight of rucksacks, commanders must carefully analyze their missions, and allow nothing but mission-essential equipment to be carried. Detailed packing lists and thorough inspection prior to a mission will insure nonessential items are not taken to the field.

When units must carry rucksacks, troops should not be expected to fight with this burden. Unit SOPS must be developed and practiced concerning the disposition of rucksacks in various situations. Rucksacks should contain nothing other than existence items, so the grounding or caching of them should have no effect on the combat effectiveness of an element or unit.

A unit moves from point to point with rucksacks on. After the movement phase of an operation is over, rucksacks should be removed and secured.

DISPOSITION OF RUCKSACKS

Techniques of handling rucksacks will vary according to the situation.

SHORT-DURATION MISSIONS (48 HOURS OR LESS)

Rucksacks should not be carried. Rations can be placed in a sock and securely tied to the back straps on the load-bearing harness. This method will not restrict mobility and noise will be reduced. The poncho, tied to the back of the pistol belt, will be the only other item needed from the existence load. The balance of the equipment to be carried, in excess of a unit's standard fighting load, will be determined by the mission. Bulky items can be carried in rucksacks, but loads should be rotated frequently.

CHANCE CONTACT

Rucksacks should be dropped immediately when contact is made, to allow rapid reaction and maneuverability. The mission to secure these rucksacks should be given to the squad or fire team that is providing rear security. The rucksacks need not be gathered or centralized in this situation, due to the usuall y short duration of this type of contact. If the unit must move because of enemy pressure, or in case of a need for reinforcements, all rucksacks must be left behind. Upon returning to retrieve dropped rucksacks, a unit must be alert for ambushes and care taken to avoid casualties from enemy boobytraps placed in and around the equipment.

PATROL BASE OPERATIONS

Elements or squads moving out of patrol bases to conduct short-duration missions should consolidate and conceal rucksacks prior to moving out. The only time rucksacks should be carried is when the mission does not call for return to the patrol base. If the patrol base is forced to withdraw from its location, all rucksacks will probably be left behind in order to break contact and to move rapidly for linkup with the balance of the unit. Reinforcements to assist a squad or element in contact while away from their patrol base should not take rucksacks when moving to their assistance. If the situation is so critical that the entire unit must move to the assistance of an element or squad, rucksacks should be quickly consolidated, concealed, and left behind.

CACHING RUCKSACKS

This technique is used as a last resort. It is a difficult mission to successfully accomplish. If a situation arises where a commander decides to cache, care must be taken in selecting a cache site. An easily identified area should be used as a reference point. From there, azimuth and pace count should be used to actual cache sites. Natural lines of drift should be avoided and care should be taken not to leave trails into cache sites. Security must be established 360 degrees around the proposed site, and must be far enough out to insure that enemy forces cannot observe the activity. Items that will be needed in the event a unit is unable to return to the cached items should be removed and included in the fighting load. (See short-duration missions.) Natural camouflage (deadfall, thickets, caves, etc.) should be used to conceal rucksacks. Pits can be dug if the area can be returned to its natural state when caching is completed. When returning to cache sites to recover rucksacks, unit leaders must consider the sites danger areas and act accordingly. Each cache site should have the equipment arranged in such a manner that it is easy to determine if anyone has tampered with the equipment. This helps in preventing casualties from boobytraps and mines. Soldiers should not group around the equipment when it is recovered and security must be maintained at all times while redistributing the rucksacks.

In all situations, the commander should remember that rucksacks are expendable. Leaders should tailor the fighting loads of their units to such a degree that the loss of all rucksacks would not hamper the unit's ability to continue the mission.

APPENDIX I

Adjustment of Indirect Fire by Sound

Section I. GENERAL

To effectively use indirect firepower, it is sometimes necessary to adjust fire, and in a jungle environment this requires special techniques. In the limited visibility of the jungle, the only method available for adjustment may be by the sound of the explosion of the round. The techniques described are equally effective in the adjustment of artillery or mortars.

The basic principle of adjustment by sound is that the direction and distance to the target are known, and the direction and distance to the explosion of the round can be calculated from the sound of the explosion. These data are compared to determine the appropriate correction in deviation and in range. The problem encountered in this method is that the heavy foliage distorts sound and makes it difficult to determine range or lateral shift corrections. There is a distinct technique used to determine the correction in each of the two dimensions.

CONTENTS

Section II. TECHNIQUES USED TO DETERMINE CORRECTIONS

CORRECTION IN DEVIATION

The direction from the observer to the target is measured from a map or by compass. The direction to the explosion is measured with a compass. These two azimuths, expressed in roils, form an angle as illustrated.

COMPUTING THE CORRECTION

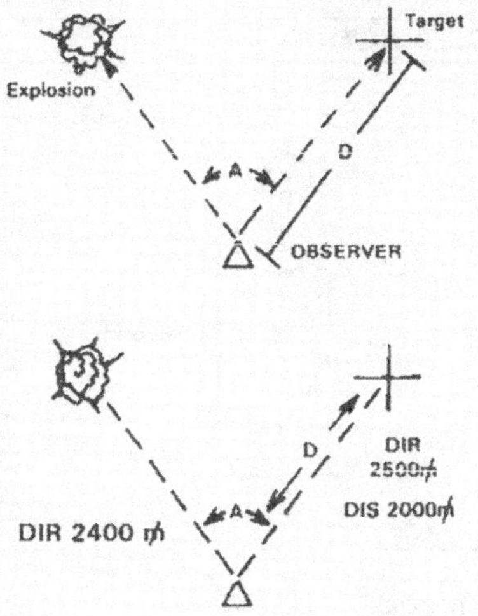

To calculate the correction in deviation, multiply angle A by distance D, with the distance expressed in kilometers, as shown in the example.

A = Angle in mils (m̸) = 2500m̸ - 2400m̸ = 100m̸

D = Distance to target in kilometers (km) = 2000m = 2 km

C = Correction in meters (m) = ?

A X D = C

100 X 2 - 200 m

The correction is right 200.

CORRECTION IN RANGE

The distance from the observer to the target can be measured on a map or can be estimated. The distance to the explosion of the round is calculated by counting the number of seconds from the impact of the projectile until the observer hears the explosion. Multiply the number of seconds by the speed of sound, 350 meters/second. Add or drop the resulting distance, as appropriate.

To tell the observer when the round impacts, the lire direction center (DFC) provides "splash and count." In splash and count, the FDC sends, "Splash - four, three, two, one-impact." On the word "impact" the round lands, and the observer starts counting the seconds until he hears the explosion.

COMPUTING THE CORRECTION

Example: After hearing "impact" the observer counted 6 seconds before hearing the explosion.

Seconds X Speed of Sound = Distance to the explosion.

6 sec x 350 m/sec = 2100m

2100m - 2000m = 100m

The correction is drop 100.

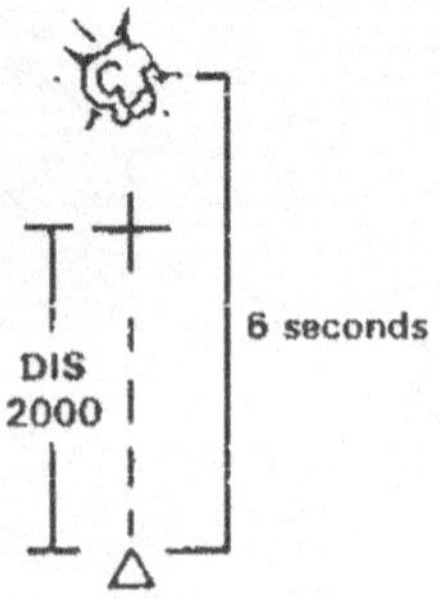

Section III. OTHER CONSIDERATIONS

The FDC normally does not send "splash and count"; rather, it must be requested in the call for fire by saying, "Cannot observe - splash and count."

Many times in dense jungle terrain, a round's effect is lost in the trees. Delay fuze can be used to penetrate the foilage and put the desired effects on the ground. A combination of fuze quick and fuze delay can be effective.

APPENDIX J

Defensive Formation

Section I. GENERAL

Jungle defensive operations are based on the same fundamentals of the defense used in other type operations. Some of the fundamentals may acquire a special significance in the jungle. The basic factors of observation and fields of fire, cover and concealment, obstacles, key terrain, and avenue of approach should always be considered carefully.

Three defensive formations will be presented here: the perimeter, the triangle, and the "Y." Platoon-size elements will be the primary consideration, although the techniques mentioned are adaptable to larger units. It should be noted that all three defensive postures have definite advantages as well as inherent disadvantages. Units should vary their type of defensive posture in order to avoid establishing a pattern.

Section II. FORMATIONS

PERIMETER DEFENSE

The first of the defensive formations is the perimeter. All elements are generally configured in a circular formation, and the terrain is used to the maximum advantage.

PERIMETER FORMATION

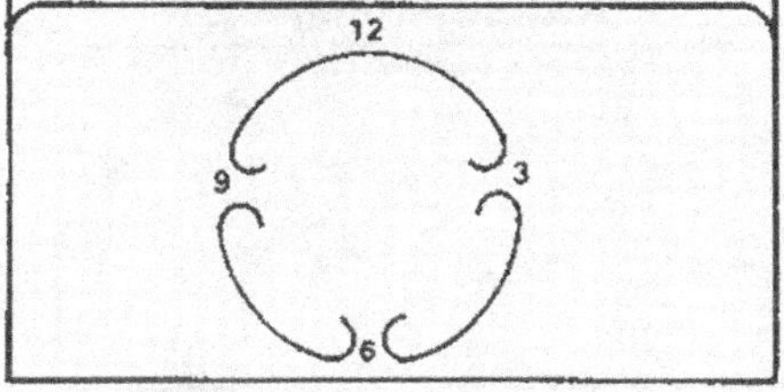

CONTENTS

Advantages are: ■ 360-degree security. ■ Centralized control. ■ Quick emplacement, and it can be executed by any size unit. ■ The frontage of the perimeter can be easily adjusted and internally reinforced.	It is mutually supported by observation/communication, and patrols are easily coordinated. Disadvantages are: ■ Any penetration creates problems of enfilade, masking, and/or cross fires. ■ It is difficult to achieve final protective fires with the crew-served weapons and, because of the circular configuration, the unit is vulnerable to the entire effects of the bursting radius of an indirect fire weapon.

The basic technique used to establish this defensive formation is the clock system. This involves the platoon using the direction of movement as 12 o'clock, with one squad occupying from 8 to 12, one squad occupying from 8 to 4, and the last squad occupying from 4 to 12. The headquarters element would normally be located in the center of the formation to facilitate control. It is generally suggested that automatic weapons be placed to cover the most likely avenues of approach. At least one machinegun should be kept with the headquarters, under the control of the unit leader, for deployment against a specific threat.

TRIANGLE DEFENSE

The second of the defensive formations is the triangle defense. This formation is a modification of the perimeter.

TRIANGLE FORMATION

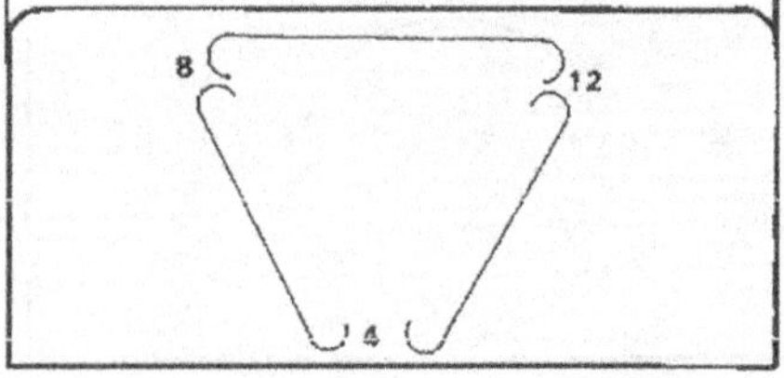

Advantages are:	Disadvantages are:
■ 360-degree security. ■ Quick emplacement, and it can be executed by any size unit. ■ A target approaching perpendicular to any side of the triangle becomes vulnerable to at least one-third of the fighting force and at least two automatic weapons. ■ It is a dual purpose formation in that it may be used as either a defensive or offensive ambush formation.	■ One or more legs of the triangle may be subjected to enfilade fire. ■ Soldiers located at the corners are bunched, thus increasing the danger from indirect fire. ■ Penetration by the enemy creates problems of enfilade, masking, and/or cross fire.

This defensive formation is best established by having the commander move forward with one element and establish a base line. Again, 12 o'clock is used as the direction of movement. The baseline could be established by the first squad, running from 8 to 12. Once this base line is established and the squad is prepared to provide support, the next squad moves forward and occupies the straight line position from 8 to 4; finally, the last squad moves forward and occupies the straight line position from 4 to 12. The headquarters element will normally locate in the center of the formation. If three machineguns are available, one will be placed at each comer in such a manner to allow flexibility in providing final protective fires down either of its two sides. If only two machineguns are available, one may be placed to cover the two sides deemed most vulnerable to attack. The second machinegun should be kept with the headquarters element under control of the unit leader. Firing positions should be pre pared at all three comers.

"Y" DEFENSE

The third defensive formation is the "Y." This formation, like the triangle, may be utilized as an offensive formation, usually in an ambush. Depending upon the situation, it can be a very effective defensive formation.

"Y" FORMATION

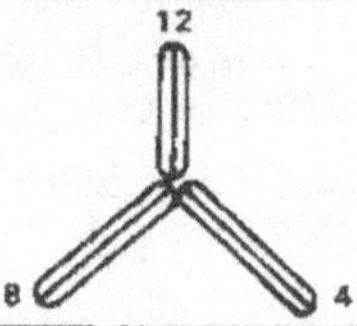

Advantages are:

- 360-degree security.
- Centralized control.
- Maximum firepower in all directions.
- Each penetration by the enemy places him in a new kill zone.

Disadvantages are:

- It must be emplaced during daylight to insure that aiming stakes are accurately emplaced, to preclude firing into adjacent positions.
- The central point of the "Y" creates bunching, and it increases the danger from indirect fire.

The recommended technique for emplacement of this formation is for the unit leader to take one element and the headquarters forward and establish the center of the sector. This establishes the 12 o'clock position. The remaining elements stay in security positions. Since all legs of the "Y" are mutually supporting, the commander has the option of having all soldiers on each leg face the same direction, or he may alternate fire teams. Once the decision is made regarding which method is to be used, subsequent legs of the "Y" must conform to the initial one. Once the lead squad is emplaced, the next squad moves forward to the center point and is closely directed into the 8 o'clock position. After the second element is in position, the third element moves forward to the center point and is closely directed into the 4 o'clock position. If three machineguns are available, one is positioned at each leg, with possible final protective fires across the fronts of adjacent legs. If only two are available, they are positioned at the center point to provide fire down the legs. This positioning of two is not preferred because it does not allow for dispersion; plus, the machineguns are bunched and extremely vulnerable to indirect fire.

Glossary

ADA	air defense artillery
ACP	armored personnel carrier
ASP	ammunition supply point
AVLB	armored vehicle launched bridge
CBR	chemical, biological, and radiological
C&C	command and control
CEOI	communications-electronics operation instructions
COMSEC	communications security
CP	command post
DZ	drop zone
FAC	forward air controller
FDC	fire direction center
FIST	fire support team
FO	forward observer
FPF	final protective fire
FRAGO	fragmentary order
FSCOORD	fire support coordinator
HE	high explosive
HEAT	high explosive antitank
HF	high frequency
ICM	improved conventional munitions
JOES	jungle operations extraction system
LAPES	low altitude parachute extraction system
LBE	load-bearing equipment
LCM	landing craft, mechanized
LZ	landing zone
METT	mission, enemy, terrain and weather, and troops and time available
MO GAS	motor gasoline
MOPP	mission oriented protection posture
NBC	nuclear, biological, chemical

OP	observation post
OPORD	operation order
ORP	objective rally point
PL	phaseline
PLL	prescribed load list
POL	petroleum, oils, and lubricants
PW	prisoner of war
PZ	pickup zone
RCLR	recoilless rifle
SLAR	side-looking airborne radar
STANO	surveillance, target acquisition, and night observation
TACAIR	tactical aircraft
TOC	tactical operations center
TOW	tube-launched, optically-tracked, wire-guided missile
VTR	vehicle track retriever

By Order of the Secretary of the Army:

16 AUGUST 1982

E. C. MEYER
General, United States Army
Chief of Staff

Official:

ROBERT M. JOYCE
Major General, United States Army
The Adjutant General

DISTRIBUTION:

Active Army and USAR: To be distributed in accordance with DA Form 12-11B, Requirements for Jungle Operations (Qty rqr block no. 307), and Operations of Army Forces in the Field (Qty rqr block no. 405).

ARNG: To be distributed in accordance with DA Form 12-11B, Requirements for Jungle Operations (Qty rqr block no. 307).

Additional copies may be requisitioned from the US Army Adjutant General Publications Center, 2800 Eastern Boulevard Baltimore, MD 21220.

☆U .S. GOVERNMENT PRINTING OFFICE : 1987 - 189-575 : QL 3

PIN: 051485-000

Made in the USA
Middletown, DE
23 July 2025

11147767R00189